Calculations for Building Craft Students

Calculations for Building Craft Students

Frank L. Tabberer, AMRSH

Hutchinson

London Melbourne Sydney Auckland Johannesburg

Hutchinson & Co. (Publishers) Ltd

An imprint of the Hutchinson Publishing Group

17-21 Conway Street, London W1P 5HL

Hutchinson Group (Australia) Pty Ltd
30-32 Cremorne Street, Richmond South, Victoria 3121
PO Box 151, Broadway, New South Wales 2007

Hutchinson Group (NZ) Ltd
32-34 View Road, PO Box 40-086, Glenfield, Auckland 10

Hutchinson Group (SA) (Pty) Ltd
PO Box 337, Bergvlei 2012, South Africa

First published 1982

© F. L. Tabberer 1982
Illustrations © F. L. Tabberer

Set in Press Roman, by Tek Art Ltd

Printed in Great Britain by The Anchor Press Ltd
and bound by Wm Brendon & Son Ltd
both of Tiptree, Essex

British Library Cataloguing in Publication Data
Tabberer, Frank L.
    Calculations for building craft students.
    1. Engineering mathematics
    510'.24624   TA331

ISBN 0 09 147201 6

Contents

Preface

Calculations for Building Craft Students replaces my previous books *Metric Calculations for Building Craft Students 1* and *2*, which were first published in 1969 and 1970. The object of this book continues to be the provision of examples and exercises in the application of basic mathematical principles to the solution of problems relating to construction processes.

The treatment in this book is essentially practical and students should make every effort to understand and complete each chapter before passing on to the next. Students should also try to associate the mathematical principles with day to day work. There is no doubt that those craftsmen who gain an aptitude in the use of calculations are the ones most likely to progress to foremanship and other supervisory positions.

Each topic is presented in three stages:

1   Explanation, to be studied carefully;
2   Examples, to be followed through step by step;
3   Exercises and tests, to be worked using the examples as patterns and/or guides.

Inevitably, the rates and prices given in the sections dealing with costs and estimating will become outdated. However, this can be the subject of a useful exercise – updating the work from the figures published regularly in the trade journals and re-working the problems.

Dimensions on drawings are, with certain exceptions, shown without unit symbols, that is:

5.850 indicates five point eight five metres; and
600 indicates six hundred millimetres.

The exceptions are:

1   When the units used are not preferred units, for example centimetres (cm).
2   When metres are in whole numbers only, for example 15 m.
3   When fractions of millimetres are used, for example 67.5 mm.
4   When mixed units are used on one drawing (see Figure 119 on page 180).

Basic SI units

Quantity	Unit	Unit symbol
Length	metre	m
Mass	kilogram	kg
Time	second	s
Electric current	ampère	A
Temperature*	kelvin	K
Luninous intensity	candela	cd

From these six units other units in the system are derived. Some of the more commonly used derived units are shown on page 8 with typical applications.

*Practical unit degree Celsius °C

Some derived SI units

Measure	Units	Application
Area	km^2 (square kilometre) ha (hectare) m^2 (square metre)	land area superficial measure generally
	mm^2 (square millimetre) 1 ha = 10000 m^2	small areas
Volume	m^3 (cubic metre) mm^3 (cubic millimetre)	cubic measure generally small volumes
Capacity	litre (no abbreviation) ml (millilitre) (1 litre = 1 dm^3)	fluid measure small measures (1 ml = 1 cm^3)
Density	kg/m^3 (kilograms per cubic metre) g/cm^3 (grams per cubic centimetre)* 1 g/cm^3 = 1000 kg/m^3	densities of building materials laboratory work

*Although the centimetre does not figure as a preferred unit of length in the construction industry, the cubic centimetre (cm^3) remains a convenient measure for small volumes and the gram per cubic centimetre (g/cm^3) is useful for experimental determination of densities of materials.

1 Definitions and basic arithmetic

A mathematical problem should not be treated as a puzzle. It is of course quite possible, by looking at the answer first, to arrive at a solution by a series of guesses, but as a craftsman you will appreciate that this method has no real value in practical work and you would be well advised to use the following hints as a guide in the solution of problems involving calculations.

1   Read each question very carefully before attempting to write anything down. Careful analysis of the conditions of a problem are essential for an accurate answer.

2   Set out your solution *neatly* and in well defined, easy-to-follow stages. You must decide from the information given the form the answer should take. The matching of these two parts of the problem is an important part of the work.

3   Always check the answer if possible. Accuracy is very important, particularly to the practical man, and a great deal of satisfaction can be derived from *knowing* that your answer is correct.

Some useful definitions

A factor

A factor of a number is any number which will exactly divide into it, for example the factors of 30 are 2, 3, 5, 6, 10 and 15.

A prime number

A prime number is a number which has no factors (except itself and one), for example 1, 2, 3, 5, 7, 11, 13, 17, etc., are all prime numbers.

Prime factors

If we divide a number into factors and then divide those factors into factors, and so on until we have only prime numbers, we then have the prime factors of the original number

for example $30 = 3 \times 10$ (these are factors of 30)
        but $10 = 2 \times 5$
    therefore $30 = 2 \times 3 \times 5$ (these are the prime factors of 30)

A composite number

A composite number is any number other than a prime number (that is to say, it has factors other than itself and one).

A common factor

A common factor is a factor *common to* two or more given numbers, for example 5 is a common factor of 25 and 75.

The highest common factor (HCF)

It is sometimes necessary to find the highest common factor of two or more numbers; that is, of course, the *largest number which will divide exactly into the given numbers*. For example, although 4 is a common factor of 16 and 24, their *highest common factor* is 8.

A multiple

A multiple is a number *into which* the given number will divide without a remainder. For example:

36 is a multiple of 12
16 is a multiple of 4

If a number is a multiple of several numbers then it is called the *common multiple* of those numbers. For example:

36 is a common multiple of 2, 3, 4, 6, 9, 12 and 18

The lowest common multiple (LCM)

A very important number in elementary calculations. It may be defined as the *lowest number* into which two or more numbers will divide exactly.

Consider the numbers 2, 4 and 6. 48 is a *high* common multiple of them. 24 is another. But we are interested only in their *lowest* common multiple which, you will find is 12.

You will probably have realized by now that a thorough knowledge of the multiplication tables is an essential part of all mathematical processes. Should you have any doubts about your own knowledge of the tables, now is the time to revise and learn them.

Example 1

Express 18, 24 and 36 in terms of their prime factors and find their HCF and LCM.

$18 = 2 \times 3 \times 3$
$24 = 2 \times 2 \times 2 \times 3$
$36 = 2 \times 2 \times 3 \times 3$

For the HCF use each prime factor which is common to all sets once only and obtain their product:

$HCF = 2 \times 3 = 6$

For the LCM use each prime factor the greatest number of times it appears in any one set:

$LCM = 2 \times 2 \times 2 \times 3 \times 3 = 72$

Example 2

Find the HCF and LCM of 21, 42 and 45.

$21 = 3 \times 7$
$42 = 2 \times 3 \times 7$
$45 = 3 \times 3 \times 5$

$HCF = 3$ (the only prime factor common to all three sets)
$LCM = 2 \times 3 \times 3 \times 5 \times 7 = 630$

Example 3

Find the HCF and LCM of 8, 27 and 36.

$8 = 2 \times 2 \times 2 \, (\times 1)$
$27 = 3 \times 3 \times 3 \, (\times 1)$
$36 = 2 \times 2 \times 3 \times 3 \, (\times 1)$

The factor 1 is not normally written but in this case it is the *only* common factor.

HCF = 1
LCM $= 2 \times 2 \times 2 \times 3 \times 3 \times 3 = 216$

The basic operations

The four basic operations in manipulating numbers are *addition, subtraction, multiplication* and *division.*

Operation	Sign	Result	Example
Addition	+	sum	$15 + 5 = 20$
Subtraction	–	difference	$15 - 5 = 10$
Multiplication	x	product	$15 \times 5 = 75$
Division	÷	quotient	$15 \div 5 = 3$

Table 1

Addition and subtraction

If unsure of working in horizontal form, that is:

$217 + 68 + 324 + 27 = 636$

the alternative vertical form is acceptable, that is:

```
 217
  68
 324
  27
 ———
 636
```

When mixed addition and subtraction is required it is best to separate the + and – numbers first and then complete the process by subtracting the – total from the + total, that is:

53 + 18 – 27+ 38 – 41 = 109 – 68 = 41

or, using the vertical form to evaluate 325 – 287 + 567 + 263 – 85:

```
 325        - 287
 567        -  85
 263        - 372
1155
-372
 783
```

Multiplication

There is more than one correct approach to long multiplication and if you are used to a particular method do not attempt to change. The method shown uses the left hand figure of the multiplier first and includes the noughts. For example:

```
(a)   254 x 127                   (b)   935 x 87
       254 (multiplicand)                935
       127 (multiplier)                   87
     254000                           74800
      50800                            6545
       1778                           81345
     306578 (product)
```

Multiplication by factors is often useful (particularly as demonstrated later in cost calculations). For example:

326 x 35

35 may be factorized as 5 x 7. Multiply first by 5, then multiply the product by 7.

```
  326
    5
 1630
    7
11410
```

Division

Division by a single digit should be done mentally from knowledge of the multiplication tables. For larger numbers long division is required. For example:

$$1961 \div 53$$

$$\text{(divisor) } 53\overline{)\,1961} \quad \begin{array}{l} 37 \text{ (quotient)} \\ \text{(dividend)} \end{array}$$

$$\underline{159} \quad (53 \times 3)$$

$$371 \quad (196 - 159, 1 \text{ brought down})$$

$$\underline{371} \quad (53 \times 7)$$

Division by factors is sometimes a quick useful method particularly if the divisor can be reduced to small factors. For example:

$$5754 \div 42$$

42 may be factorized as 2 x 3 x 7. Divide first by 2, then divide the quotient by 3 and finally divide the second quotient by 7 to obtain the final quotient.

$$2\underline{)\,5754}$$
$$3\underline{)\,2877}$$
$$7\underline{)\,959}$$
$$\quad\;\;137$$

Various aids to calculation make it unnecessary to carry out long and tedious operations. The electronic calculator is the quickest. However, one should not fall into the bad habit of depending entirely on a calculator - the brain needs exercise!

Exercise 1

1   Find as many sets of factors as you can for each of the following numbers:

(a) 36   (f) 45
(b) 91   (g) 100
(c) 128   (h) 121
(d) 211   (i) 175
(e) 324   (j) 295

2   Which of the following are prime numbers?
(a) 43
(b) 83
(c) 97
(d) 133
(e) 171

3   What are the prime factors of the following numbers?
(a) 27
(b) 28
(c) 84
(d) 144
(e) 217

4   Find the HCF of each of the following groups of numbers:
(a)   8, 20, 24 and 36
(b)   14, 49, 56 and 63
(c)   28, 70, 84 and 126

5   Find the LCM of each of the following groups of numbers:
(a)   3, 6, 8 and 12
(b)   7, 21, 28 and 42
(c)   6, 8, 16 and 24

6   Find the HCF and the LCM of each of the following groups of numbers:
(a)   4, 8, 12 and 18
(b)   6, 9, 15 and 18
(c)   15, 27 and 135

7   Find the sum of 2548, 3876, 1714 and 894.

8   Add 19,328, 4603, 456 and 937.

9   Subtract 693 from 2471.

10   Find the difference between 84 and 1738.

11   Multiply 2794 by 461.

12   Find the product of 83, 59 and 142.

13   Divide 7812 by 63.

14   How many times is 87 contained in 18,705?

Mixed operations

The evaluation of expressions containing multiplication and/or division signs as well as plus and/or minus signs requires the application of certain rules of precedence. Failure to do so can lead to serious error. The order of precedence is:

1   Carry out multiplication and division;
2   Carry out addition and subtraction.

This is demonstrated in the following examples.

$$17 \times 6 + 33 \times 8 = 102 + 264$$
$$= 366$$

It would be wrong to add the 6 and 33 together first to obtain 17 × 39 × 8. This would give a result of 5304. This would have been the result if the original expression had been given as 17 × (6 + 33) × 8, the brackets indicating that the 6 and 33 should be taken together as a combined value.

$$963 \div 9 - 121 \div 11 \; = 107 - 11$$
$$= \; 96$$
$$123 \times 5 + 183 \div 3 - 75 \times 6 \; = 615 + 61 - 450$$
$$= 676 - 450$$
$$= 226$$

Brackets

As indicated above brackets are used to group a number of terms together, the group to be treated as one quantity. This means that the operations inside the brackets must be carried out first, taking overall precedence.

$$26 \div (32 - 19) \times (2 + 18 - 13) \; = 26 \div 13 \times 7$$
$$= 2 \times 7$$
$$= 14$$

Averages

When a number of quantities vary slightly it is often convenient to use their average value. This is the value each would have if they were all equal and their total remained as before. To find the average add the terms and divide their sum by the number of them, for example, the average of 455, 463, 443, 452, 449 and 462 is:

$$(455 + 463 + 443 + 452 + 449 + 462) \div 6 \; = 2724 \div 6$$
$$= 454$$

Since all the terms are between 400 and 500 an alternative method is:

$$400 + \frac{55 + 63 + 43 + 52 + 49 + 62}{6} = 400 + \frac{324}{6}$$
$$= 400 + 54$$
$$= 454$$

Exercise 2

1   Evaluate the following expressions:
   (a) $8 \times 55 - 7 \times 32$
   (b) $108 \div 27 + 116 \times 11$
   (c) $15 \times 17 + 37 \times 18 - 368 \div 23$
   (d) $33 + 72 \div 9 - 245 \div 7$
   (e) $910 - 16 \times 43 + 185 \div 37$

(f)   12 x (14 + 11 + 36 – 25)

(g)   22 x (36 – 18) x (13 – 5) ÷ (128 – 96)

(h)   (14 + 37 – 21) ÷ (56 – 26)

(i)   11 x [15 + 3(25 – 18)] (inner brackets first)

(j)   [876 – 23(213 – 179)] ÷ (18 + 29)

2   Find the averages of the following:

(a)   68, 93, 72, 27

(b)   161, 175, 171, 159, 164

(c)   27, 39, 46, 45, 37, 22

(d)   2964, 2952, 2971, 2958, 2969, 2970

2 Fractions

Now that we use the decimal system the use of vulgar fractions should gradually lessen. However, using metric weights and measures does not mean that we should dispense with vulgar fractions altogether, in fact we still use the $\frac{1}{2}$ pence in coinage. You should understand the meaning of fractions and be able to manipulate them, particularly their conversion to decimal form.

If you cut a brick into two equal parts, each part will be one half of the brick; this may be written '$\frac{1}{2}$'.

So, one half = $\frac{1}{2}$ (this is a fraction).

Take a sheet of lead and cut it into four equal parts; each part will represent one quarter; this may be written '$\frac{1}{4}$' (this is a fraction).

Cut a piece of timber into six equal parts; each part is one sixth ($\frac{1}{6}$) of the whole. Take five of these parts and you now have five sixths of the whole; this may be written $\frac{5}{6}$ (this is a fraction).

Notice that in each case the *lower part* of the fraction indicates the number of parts into which the whole unit is divided; this is called the *denominator*.

Notice also that the *upper part* of the fraction indicates the number of parts being used; this is called the *numerator*. Thus:

$\frac{7 \text{ Numerator}}{8 \text{ Denominator}}$ means that the whole unit is divided into 8 parts and 7 of them are being used

The line between the numerator and the denominator is called the *bar* of the fraction.

Fractions written in this form are called *vulgar fractions* (as opposed to *decimal fractions* which we shall meet later).

Types of vulgar fraction

If the numerator is *smaller than* the denominator then the fraction is called a *proper fraction*. For example:

$\frac{3}{4}, \frac{5}{16}, \frac{11}{12}, \frac{15}{16}$ are all proper fractions

Each fraction is less than 1 and therefore truly a *part* of a whole one. Hence the term *proper*.

If, however, the numerator is *greater than* the denominator then the fraction is called an *improper fraction*. For example:

$\frac{5}{4}, \frac{18}{12}, \frac{27}{16}$ are all improper fractions

Obviously each fraction is greater than 1 so it cannot be a *part* of a whole one. Hence the term *improper*. In this case we may obtain what is called a *mixed number* by dividing the denominator into the numerator and writing down the remainder as a proper fraction. For example:

(a) $\frac{5}{4} = 1\frac{1}{4}$

(b) $\frac{18}{12} = 1\frac{\overset{1}{\cancel{6}}}{\underset{2}{\cancel{12}}} = 1\frac{1}{2}$

Notice in example (b) the method of cancellation used to reduce the proper fraction to its lowest terms. Cancellation is the division of both the numerator and the denominator by a common factor, preferably their HCF.

Many simple problems may be solved by the use of fractions, provided the following methods are clearly understood:

Adding and subtracting fractions

1   Collect all whole numbers.

2   Reduce all the fractions to a common denominator. This will be the LCM of all the denominators.

3   Collect the numerators and unite the final fraction with the whole number already found, finally reducing the fraction to its lowest terms.

Follow these stages carefully in the examples shown.

Example 1

Simplify: $4 + 6\frac{1}{8} + 3\frac{2}{3} - 5\frac{1}{4}$

Expression $= 8\dfrac{3 + 16 - 6}{24}$

$= 8\frac{13}{24}$

Note: In the second line, 24, being the LCM of 8, 3 and 4, becomes the common denominator. The fractions are converted to $\frac{1}{24}$ths and a common bar used for them all.

Example 2

Simplify: $5\frac{3}{8} + 2\frac{1}{8} - 3\frac{1}{4}$

$$\text{Expression} = 4\frac{3 + 1 - 2}{8}$$
$$= 4\frac{2}{8}$$
$$= 4\frac{1}{4}$$

Example 3

Simplify: $4\frac{3}{5} - 3\frac{1}{15} + 1\frac{7}{10}$

$$\text{Expression} = 2\frac{18 - 2 + 21}{30}$$
$$= 2\frac{37}{30}$$
$$= 3\frac{7}{30}$$

Note: The numbers must be taken progressively, that is, dealing with the whole numbers, the 3 is subtracted from the 4 and then the 1 is added to the result, giving 2. It would have been wrong to add the 3 and the 1 together first and then take the result from 4. The same applies to numerators of the fractions.

Example 4

Simplify: $3\frac{4}{9} - 2\frac{2}{3}$

$$\text{Expression} = 1\frac{4 - 6}{9}$$

Since 6 cannot be subtracted from 4 the 1 must be changed to $\frac{9}{9}$ths and combined with the $\frac{4}{9}$:

$$= \frac{13 - 6}{9}$$
$$= \frac{7}{9}$$

Example 5

Simplify: $\frac{6}{7} + 17 + \frac{3}{14} - (4 + 6\frac{1}{2})$

Notice the *brackets* here, which indicate that this section must be simplifed *first*.

$$\text{Expression} = \frac{6}{7} + 17 + \frac{3}{14} - 10\frac{1}{2}$$
$$= 7\frac{12 + 3 - 7}{14}$$
$$= 7\frac{8}{14}$$
$$= 7\frac{4}{7}$$

Multiplying and dividing fractions

When numbers involving fractions are to be multiplied, the following procedure is adopted:

1 Change all mixed numbers to improper fractions.

2 Perform any cancelling that is possible.

3 Multiply all the numerators together to give a new numerator, and all the denominators together to give a new denominator.

4 If this new fraction is an improper fraction change it to a mixed number.

Example 6

Simplify: $4\frac{1}{3} \times \frac{1}{2} \times 4\frac{1}{5}$

$$\text{Expression} = \frac{13}{\cancel{3}_{1}} \times \frac{1}{2} \times \frac{\cancel{21}^{7}}{5} \text{ (cancelling by 3)}$$

$$= \frac{91}{10}$$

$$= 9\frac{1}{10}$$

Example 7

Find $\frac{4}{5}$ of $18\frac{1}{2}$

Note: The word *of* means multiply.

$$\text{Expression} = \frac{4}{5} \times 18\frac{1}{2}$$

$$= \frac{4}{5} \times \frac{37}{\cancel{2}_{1}}^{2} \text{ (cancelling by 2)}$$

$$= \frac{74}{5}$$

$$= 14\frac{4}{5}$$

When a fraction is to be *divided* by another fraction, first bring both to improper fractions, then *invert the divisor* (the fraction you are dividing by) and *multiply*.

Example 8

Simplify: $6\frac{1}{4} \div 1\frac{1}{2}$ $\left(\text{this could have been written } \dfrac{6\frac{1}{4}}{1\frac{1}{2}} \right)$

Expression $= \frac{25}{4} \div \frac{3}{2}$

$= \frac{25}{\underset{2}{4}} \times \frac{\overset{1}{2}}{3}$ (cancelling by 2)

$= \frac{25}{6}$

$= 4\frac{1}{6}$

Rules of precedence

When a number of different operations are required to simplify an expression certain rules of precedence must be observed. This has already been mentioned in Chapter 1 and the rules are given here in more detail.

They may be summarized as follows.

1   Work out the contents of brackets.
2   Change 'of' to multiply.
3   Invert divisors and change to multipliers.
4   Carry out multiplication.
5   Carry out addition and subtraction.
6   Simplify the result if necessary.

Example 9

$2\frac{2}{3}(6\frac{1}{4} - 3\frac{1}{2}) + 2\frac{7}{8}$

$= 2\frac{2}{3}(3\frac{1-2}{4}) + 2\frac{7}{8}$

$= 2\frac{2}{3} \times 2\frac{3}{4} + 2\frac{7}{8}$

$= \frac{8}{3} \times \frac{11}{4} + 2\frac{7}{8}$

$= \frac{22}{3} + 2\frac{7}{8}$

$= 7\frac{1}{3} + 2\frac{7}{8}$

$= 9\frac{8+21}{24}$

$= 9\frac{29}{24}$

$= 10\frac{5}{24}$

Example 10

$[(5\frac{1}{4} - 2\frac{3}{8}) \div 1\frac{1}{2}] + [3\frac{1}{4} \div (\frac{2}{3} + 1\frac{1}{5})]$

$= \left[\left(3\frac{2-3}{8}\right) \div 1\frac{1}{2}\right] + \left[3\frac{1}{4} \div \left(1\frac{10+3}{15}\right)\right]$

$$= (2\tfrac{7}{8} \div 1\tfrac{1}{2}) + (3\tfrac{1}{4} \div 1\tfrac{11}{15})$$
$$= \left(\frac{23}{8} \times \frac{2}{3}\right) + \left(\frac{13}{4} \times \frac{15}{26}\right)$$
$$= \frac{23}{12} + \frac{15}{8}$$
$$= 1\tfrac{11}{12} + 1\tfrac{7}{8}$$
$$= 2\frac{22 + 21}{24}$$
$$= 2\tfrac{43}{24}$$
$$= 3\tfrac{19}{24}$$

Exercise 3

1   Write down three examples of each of the following:
   (a)   a proper fraction
   (b)   an improper fraction
   (c)   a mixed number

2   Convert the following to improper fractions:
   (a) $2\tfrac{1}{2}$
   (b) $4\tfrac{1}{5}$
   (c) $19\tfrac{7}{16}$
   (d) $3\tfrac{3}{8}$
   (e) $23\tfrac{1}{4}$

3   Convert the following to mixed numbers:
   (a) $\tfrac{17}{5}$
   (b) $\tfrac{23}{7}$
   (c) $\tfrac{46}{21}$
   (d) $\tfrac{49}{11}$
   (e) $\tfrac{117}{20}$

4   Evaluate the following:
   (a) $4\tfrac{1}{12} + 7\tfrac{1}{4} - 3\tfrac{1}{16} + 2\tfrac{1}{8}$
   (b) $3\tfrac{3}{4} + \tfrac{7}{16} - 1\tfrac{5}{8}$
   (c) $2\tfrac{6}{7} - 4\tfrac{3}{4} + 2\tfrac{3}{14}$
   (d) $7\tfrac{31}{64} - 2\tfrac{3}{16} - 1\tfrac{7}{8} - 1\tfrac{15}{32}$

5   Evaluate:
   (a) $1\tfrac{2}{3}$ of $45\tfrac{1}{2}$
   (b) $\tfrac{8}{11}$ of $(4\tfrac{1}{8} \times \tfrac{1}{4})$
   (c) $4\tfrac{1}{2} - 1\tfrac{1}{6}$
   (d) $6\tfrac{1}{2} \times 2\tfrac{3}{8} - 11\tfrac{1}{4}$

6   Simplify and evaluate:
   (a) $2\frac{3}{5} \times 1\frac{1}{4} \div (1\frac{1}{2} + 2\frac{1}{3})$
   (b) $(4\frac{2}{5} \div 2\frac{9}{20}) + (3\frac{1}{5} \div 2\frac{2}{3})$
   (c) $4\frac{1}{2} \times 1\frac{2}{3} - \frac{6}{7} \text{ of } 2\frac{4}{5}$
   (d) $[(\frac{15}{16} - \frac{1}{3}) \times 1\frac{3}{4}] - \frac{7}{12}$

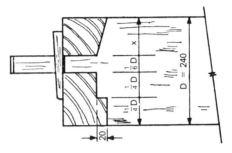

Figure 1

7   Figure 1 shows a method of setting out a tusk tennon joint. Calculate the actual dimensions of the joint adding the missing dimension marked 'x'.

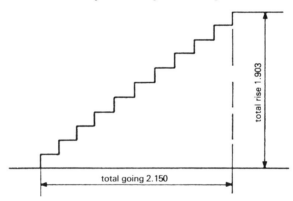

Figure 2

8   Calculate the rise and going of each step from the illustration shown in Figure 2.

9   A water tank can be filled from its inlet pipe in 30 minutes and emptied from its outlet pipe in 45 minutes. What fraction of the tank is filled in one minute with both pipes open?

3 Metric systems – SI units

So far we have used mainly numbers and not quantities, that is, we have not been using *units*. Ever since man began to trade, build and manufacture, various systems of units have been used. In Britain and countries associated with Britain the imperial system has been long established. It uses a complexity of multiples and sub-multiples such as 2240 pounds to a ton, 16 ounces to a pound, 12 inches to a foot, 3 feet to a yard, 1760 yards to a mile and so on.

Metric systems are those based on the metre as the unit of length and are decimal systems, so that all multiples and sub-multiples of units are powers of ten. There is more than one metric system. In the one known as the MKS system quantities are measured mainly on the basis of three units, metre for length, kilogram for mass and second for time. The CGS system has as its basic units the centimetre, gram and second. In these systems there are many derived units which are units built up from the basic units for measuring those values which involve more than one quantity. Examples are density, pressure, energy, power and so on. The development of science and technology in modern times has resulted in many new derived units being established. Lack of complete international control has resulted in some confusion and to rectify this situation a committee was established which, after some years, developed a system acceptable to all countries. This is known as the Système International d'Unités, abbreviated to the SI system.

This SI system has been adopted in Britain. You will find details of some of the units on pages 7 and 8, together with their main multiples and sub-multiples. Table 2 lists a large range of standard symbols of metric multiples and sub-multiples (although this is not a complete list). You will not require to use them all in your present work but they are worth noting for future reference. The multiple prefixes are Greek and the sub-multiple prefixes are Latin.

In the following worked examples and throughout this book SI units and their multiples and sub-multiples are used.

Decimal fractions may be regarded as vulgar fractions which have 10, 100, 1000, etc., as their denominators.

$\frac{5}{10}$, $\frac{63}{100}$, $\frac{459}{1000}$ are all vulgar fractions which may be converted directly to decimal fractions by omitting the denominators and placing decimal points in front of the numerators.

Multiplication factor		Prefix	Symbol
1 000 000	10^6	mega	M
1 000	10^3	kilo	k
100	10^2	hecto	h
10	10^1	deca	da
0.1	10^{-1}	deci	d
0.01	10^{-2}	centi	c
0.001	10^{-3}	milli	m
0.000001	10^{-6}	micro	μ

Table 2

Thus, $\frac{5}{10}$ = .5; $\frac{63}{100}$ = .63; $\frac{459}{1000}$ = .459.

Such values are usually written with a nought before the decimal point. That is:

0.5, 0.63, 0.459

This avoids any doubt about the position of the decimal point.

The figures to the left of the decimal point indicate the whole number. The figures to the right of the point indicate the fraction, the first column containing tenths, the second hundredths, the third thousandths, and so on. For example:

(a) $3\frac{5}{10}$ = 3.5   (three point five)

(b) $49\frac{53}{100}$ = 49.53 (forty-nine point five three. Do not make the mistake of calling this forty-nine point fifty-three)

(c) $5\frac{13}{1000}$ = 5.013 (five point nought one three)

Changing vulgar fractions to decimals

This is achieved by dividing the denominator into the numerator thus:

(a) $\frac{7}{8}$ = 0.875 because 8) 7.000
                              0.875

(b) $\frac{3}{5}$ = 0.6   because 5) 3.0
                              0.6

(c) $13\frac{3}{4}$ = 13.75 because 4) 3.00
                              0.75

Note: When dealing with mixed numbers as in (c) do not include the whole number part in your division.

Extended decimal fractions

Many vulgar fractions make lengthy decimals when converted; in fact some are unending! In such cases it is usual to evaluate only to the degree of accuracy required by the work concerned.

You may be asked to give an answer correct to 'so many' places of decimals. Proceed as follows:

Method

1   Carry out the division to one more place of decimals than is required.

2   If the last figure is 5 or over, add one to the preceding figure.

3   If the last figure is less than 5 discard it completely.

Example 1

Express $\frac{2}{7}$ as a decimal correct to three places of decimals.

Dividing 2 by 7, we have 7)2.00
                             0.2857

Since the fourth figure is over 5, the third figure becomes 6 and the answer is 0.286 *correct to three places of decimals.*

Example 2

Express $50\frac{1}{9}$ as a decimal correct to two places of decimals.

Dividing 1 by 9, we have 9)1.000
                             0.111

Since the third figure is under 5 the answer correct to two places of decimals is 50.11, a recurring decimal. If the division had been continued the result would have been 50.111111111 ..., etc. It is written as 50.$\dot{1}$ with a dot over the recurring figure.

Significant figures

Values are sometimes required to 'so many' significant figures. The procedure is then as before, but the number of figures in the answer are counted *including the whole number part.*

Example 3

Express 1.0346 correct to four significant figures.
   There are 5 figures altogether in the number. As the last figure is over 5 the preceding figure must be increased to 5 and the answer is 1.035 *correct to four significant figures*.

Example 4

Express 123.73 correct to four significant figures.
   Since the last figure is less than 5 it is discarded and the answer is 123.7 *correct to four significant figures.*

Degree of accuracy

This is largely a matter of common sense. You would not express the area of a plot of land as 450.001 m^2 as this is only 10 cm^2 over the 450 m^2, a negligible amount. Again you would not work out the capacity of a large reservoir tank to the nearest millilitre (ml) since this is an extremely small unit. On the other hand you would not give a person's wages as £12.60 instead of £12.59 as the difference is 1 pence and this sort of approximation could lead to considerable losses on a large wages bill.

Calculations involving decimals

Addition and subtraction

This should not present any difficulty provided the decimal points are kept directly underneath one another. Failure to do this will always give you trouble and probably lead to errors. For example:

(a)   563.624   (b)   2.45
      -247.324   3.271
      -------   +16.008
      316.300   -------
                              21.729

Example 5

What is the inside diameter of the cylinder shown in Figure 3 which has an outside diameter of 1.240 m and is made of concrete 75 mm thick?

Inside diameter = 1.240 – (2 x 0.075) (changing the 75 mm to m)
            = 1.240 – 0.150
            = 1.090 m

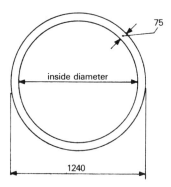

Figure 3

Multiplying decimals

The problem of placing the decimal point correctly is easily solved if you remember that the number of decimal places in the answer will always be the sum of the number of decimal places in the two numbers which you have multiplied.

Note: The noughts, if any, must also be counted when applying this rule.

Example 6

16.342 x 7.32

```
      16.342
       7.32
    114 394
      4 9026
       32684
    119 62344
```

Now, fix the position of the decimal point in the answer as follows:

1   Count the total number of decimal places in the question – in this case five, made up of three in 16.432 and two in 7.32.

2   Transfer this number of decimal places to the answer by counting from *right* to *left*. In this case a total of 5 decimal places gives an answer of 119.62344.

Example 7

Multiply 106.72 by 23.4 and give the answer to six significant figures.

```
    106.72
     23.4
 2134 4
  320 16
   42 688
 2497 248 = 2497.248
           = 2497.25
```

Example 8

Find by using decimals the value of $1\frac{3}{5}$ of $21\frac{3}{8}$. (First express mixed numbers as decimals.)

$21\frac{3}{8} = 21.375$ $1\frac{3}{5} = 1.6$

```
    21.375
      1.6
 21 375
 12 8250
 34 2000  = 34.2000
          = 34.2
```

Note: The noughts must be included when finding the position of the decimal point.

Division of decimals

In this it is necessary to make the *divisor* (the number you are dividing by) into a whole number by moving the point the required number of places to the right. Move the point the same number of places to the right in the *dividend* (the number you are dividing into), and then proceed with the division. For example:

Divide 16.32 by 0.2

This may be written $\dfrac{16.32}{0.2}$ and becomes $\dfrac{163.2}{2}$.

Now both numerator and denominator are ten times as large so that the value of the fraction remains the same. The answer is obviously 81.6.

Example 9

Find the value of $\dfrac{16.75}{0.25}$

This becomes $\dfrac{1675}{25}$ =

$$
\begin{array}{r}
67 \\
25\overline{)1675} \\
150 \\
\hline
175 \\
175 \\
\hline
\cdots
\end{array}
$$

$$= 67$$

Example 10

Divide 9.375 by 4.76 and give the answer to three significant figures:

$$\frac{9.375}{4.76} = \frac{937.5}{476} =$$

$$
\begin{array}{r}
1.969 \\
476\overline{)937.500} \\
476 \\
\hline
461.5 \\
428.4 \\
\hline
33.10 \\
28.56 \\
\hline
4.540 \\
4.284 \\
\hline
.256
\end{array}
$$

= 1.97 to three significant figures

Study the following examples carefully before attempting the final exercise in this chapter.

Example 11

Find the value of $\dfrac{(2.25 + 14.6) \times (3.4 - 1.15)}{0.05}$

Remembering to work the brackets first:

$$
\begin{array}{ll}
2.25 & 3.40 \\
+14.60 & -1.15 \\
\hline
16.85 & 2.25
\end{array}
$$

Fraction $= \dfrac{16.85 \times 2.25}{0.05}$

Now the multiplication:

$$
\begin{array}{r}
16.85 \\
2.25 \\
\hline
8425 \\
3\,370 \\
33\,70 \\
\hline
37\,9125
\end{array}
$$

Fraction $= \dfrac{37.9125}{0.05}$

$= \dfrac{3791.25}{5}$

Dividing denominator into numerator = 758.25

Example 12

One litre of fresh water weighs approximately one kilogram. Find the volume of 560 grams of water.

1 kg of water occupies 1 litre
1 g of water occupies 1 ml
(since 1 g is a thousandth of a kg and 1 ml is a thousandth of a litre)

Therefore, 560 g occupies 560 ml.

Example 13

Two hundred and fifty steel frame-ties are to be made of flat bar. The steel weighs 0.387 kg per metre run and each tie is 229 m long. Find the total weight of steel required.

Total length required to make 250 ties
= 250 x 0.229 m (converting the 229 mm to m)
= 57.250 m

Total weight at 0.387 kg/m
= 57.250 x 0.387 kg
= 22.156 kg

Exercise 4

1   Convert to decimals:

(a) $21\frac{7}{10}$ (d) $\frac{113}{1000}$ (g) $\frac{27}{10}$

(b) $14\frac{167}{1000}$ (e) $\frac{1}{1000}$ (h) $\frac{471}{100}$

(c) $2\frac{9}{10}$ (f) $3\frac{19}{100}$

2   Convert to decimals, correct to three places if extended:

(a) $\frac{21}{64}$ (d) $\frac{1}{4}$

(b) $\frac{2}{3}$ (e) $\frac{7}{16}$

(c) $\frac{5}{8}$ (f) $14\frac{1}{12}$ to four significant figures

(g) $21\frac{1}{5}$ (i) $101\frac{2}{15}$

(h) $3\frac{3}{4}$ (j) $\frac{32}{625}$

3   Find the value of each of the following:
- (a) $8.706 + 0.024 + 13.301 - 5.045$
- (b) 3.64×0.0037 to three significant figures
- (c) 0.02×1.6
- (d) $1.4 \div 0.7$
- (e) $\dfrac{3.5 \times 1.2}{0.07}$
- (f) $1.5 \div 0.015$

4   Find the value of:

$$\frac{(3.2 + 0.004 - 1.111) \times 0.25}{(4 \div 0.2) - 17.907}$$

5   Calculate the number of m^3 contained in 0.632 of a Petrograd standard of timber if one standard contains 4.670 m^3.

6   If the average time allowed for a plumber and apprentice to fix a lavatory basin complete, including joints to pipes, is 2.55 hours, what would be the estimated labour cost of fixing six such basins? Rates: plumber 203 pence per hour; mate 150 pence per hour.

7   If the materials required for 1.5 m^3 of cement mortar are: cement 560 kg; sand 1.5 m^3; how much cement and sand will be required for a total of 10 m^3 of cement mortar?

8   The weight per square metre of lead 3 mm thick is 34.230 kg. What weight of lead is required to cover a roof measuring 5.500 m x 3.250 m?

9   The number of rolls of wallpaper required to cover a certain area
$$= \frac{\text{area to be covered in m}^2}{\text{area of one roll in m}^2}.$$
If the area of one roll of paper is 3.75 m^2, how many rolls will be required to cover a total of 40 m^2?

10   The capacity of a water cistern in litres equals:
length (m) x breadth (m) x depth (m) x 1000

What is the capacity of each of the following cisterns?
- (a)   length 610 mm, breadth 430 mm, depth 430 mm.
- (b)   length 610 mm, breadth 460 mm, depth 480 mm.

Answer to the nearest litre.

4 Percentage, ratio and proportion

Percentages

This is an important section with many practical applications and should be studied carefully.

The words *per cent* mean *per hundred*, and the symbol % is used to denote 'per cent'. Thus:

10 per cent, or 10%, means $\frac{10}{100}$ ($\frac{1}{10}$ or 0.1)

5 per cent, or 5%, means $\frac{5}{100}$ ($\frac{1}{20}$ or 0.05)

25 per cent, or 25%, means $\frac{25}{100}$ ($\frac{1}{4}$ or 0.25)

Note: Any fraction may be converted to a percentage by multiplying it by 100 – always provided the % sign follows the result. For example:

Convert $\frac{1}{8}$ to a percentage

Percentage $= \frac{1}{8} \times \frac{100}{1} = \frac{25}{2} = 12\frac{1}{2}\%$

Many fractions may be converted to percentages (and vice versa) mentally.

Try the first two questions in Exercise 5 on page 39.

To increase a quantity by, say, 15%, that is $\frac{15}{100}$, multiply the quantity by $\frac{115}{100}$.

That is $\frac{100}{100}$ for the original quantity plus $\frac{15}{100}$ for the increase, $\frac{100}{100} + \frac{15}{100} = \frac{115}{100}$.

To increase 5450 by 10%

$\frac{5450}{1} \times \frac{110}{100} = 5995$

Basic applications

Discounts and list prices

The prices of building materials and components are often subject to *trade discount*: this means that you may be given the privilege of paying a certain percentage less than the advertised price for the goods. Follow the example carefully.

Example 1

Here is a quotation from a builder's merchant.

1   5000 bricks @ £145.25 per 1000 less trade discount of 10%
2   5 m³ sand @ £20.25 per m³ less trade discount of 5%

The net cost (gross cost less discount) of each item is calculated as follows:

1   *Bricks*
   5000 @ £145.25 per 1000 = £145.25 x 5   = £726.250
   Discount (10% = $\frac{1}{10}$)   = £ 72.625
                                   Net cost   = £653.625

This should be written £653.62$\frac{1}{2}$

2   *Sand*
   5 m³ @ £20.25 = £20.25 x 5   = £101.2500
   Discount (5% = $\frac{1}{20}$)   = £   5.0625
                      Net cost   = £ 96.1875

This, to the nearest penny, is £96.19

Bonuses

You may be fortunate enough to receive in your pay packet a bonus of, say, 5%. This means that your pay will be *increased* by $\frac{5}{100}$ or $\frac{1}{20}$.
   Take the following example:
   An apprentice receives a bonus of 5%; how much does he receive if his normal wage is £49.80?
   One way of calculating this is to divide by 100 by moving the decimal point two places to the left (£0.498). This is 1%.
   Now multiply it by 5:

£0.498 x 5 = £2.49

   Amount received = £49.80 + £2.49 = £52.29

Wastage of materials

Wastage may be kept to a minimum by careful handling (and this is where you as a craftsman can help). It is, however, likely that some damage will occur and, of course, wastage due to cutting is unavoidable. A percentage allowance for this is usually made when estimating.

Example 2

A certain job is estimated to require a total of 10,650 bricks, plus an allowance of 10% for cutting and waste. How many bricks should be ordered?

$$\text{Total required} = \frac{1065\emptyset}{1} \times \frac{11\emptyset}{1\emptyset\emptyset}$$
$$= 11,715 \text{ bricks}$$

Overheads

The overheads of any business or firm are the running costs relating to that business, and in pricing a job it is usual to add a percentage of the total cost to allow for these. In addition the firm should make some profit. Overheads are sometimes linked with profit.

Example 3

If the actual cost of labour and materials for a job amounts to £2165, give a price for the job allowing 20% for overheads and profit.

$$\text{Total price} = \frac{\overset{433}{2\cancel{16\emptyset}}}{1} \times \frac{\overset{6}{12\emptyset}}{\underset{5}{1\emptyset\emptyset}}$$
$$= £2598$$

Increase in bulk (excavations)

When earth is excavated it will *increase in bulk* (so that you may find yourself apparently carting away more earth than has been excavated). This, of course, is due to the fact that the earth is far more loosely packed *after* excavation, so that it occupies a greater volume.

In a similar way, a given amount of sand will change in bulk according to the amount of moisture present. This is quite important when calculating quantities of materials for mortar and concrete, etc. Percentage allowances are usually made for these and similar problems.

Example 4

How many lorry loads will be required to cart away clay soil from an excavation totalling 320 m^3, if the increase in bulk is estimated to be 25% and one lorry will take 5 m^3 per load?

1   Total excavated volume $= 320 \text{ m}^3 + 25\% \text{ of } 320 \text{ m}^3$
    25% $(\frac{1}{4})$ of 320 $=\ \ 80 \text{ m}^3$
    Required volume $= 400 \text{ m}^3$

2   Number of loads $= \dfrac{400}{5}$

    $=\ \ 80$

Simple interest

You may find it necessary to borrow money for a period. In such cases a *percentage interest* may be charged. The original amount of money is called the *principal,* and the *interest* is usually charged at a certain percentage per annum, called the *rate* of interest. The *amount* outstanding is equal to the principal plus the interest at any given time.

Example 5

Find the *amount* and the *interest* of £150 at 5% per annum for a period of 3 years.
To find the interest, multiply:

principal x rate of interest x number of years

$$= \frac{15\cancel{0}}{1} \times \frac{\cancel{5}}{\underset{2}{\cancel{100}}} \times \frac{3}{1}$$

$= £22.50$
Amount $= £150 + £22.50$
$= £172.50$

Summing up

Note the following facts which will help you to simplify certain problems:

1   A *bonus* of 5% means 5p in the £ because there are 100p to £1.

2   A *discount* of 15% on a purchase means a reduction in cost of 15p in the £.

3   To allow 10% for *wastage* means to add on $\frac{1}{10}$ of the quantity apparently required.

4   A *profit* of 20% means that you have made a profit of $\frac{1}{5}$ on the cost of the job, because $20\% = \frac{20}{100} = \frac{1}{5}$ of the cost.

5   An *increase in bulk* of 25% means that the material will increase in volume by $\frac{1}{4}$, because $25\% = \frac{25}{100} = \frac{1}{4}$ increase in volume.

Ratio and proportion

There are several ways of giving the relation of one quantity to another; a very convenient method of comparison is to state the *ratio* between them. Thus:

If £24 is to be divided between 2 men in the ratio of one to five, it means that one man will get £1 as often as the other gets £5. This ratio may be written as 1:5, and indicates that of every *six parts* (1 + 5), one man will get one part and the other man five parts.

One man receives $\frac{1}{6}$ of £24   = £4
The other man receives $\frac{5}{6}$ of £24 = £20

Note the expressions
in the ratio of
proportionally
in the same proportion
pro rata
which have practically the same meaning.

Study the following examples carefully.

Example 6

If cement mortar is mixed in the proportion of 1:3 parts Portland cement and sand by volume, how much cement and sand will be required to make 28 m^3 of mortar?

Total number of parts will be 3 + 1 = 4, of which 1 will be cement and 3 sand. Thus:

$\frac{1}{4}$ of 28 = 7 m^3 of cement
$\frac{3}{4}$ of 28 = 21 m^3 of sand

(Actually this problem is not a very practical one, since if 21 m^3 of sand were mixed with 7 m^3 of cement, the result would be considerably less than 28 m^3 of mortar. This aspect will be dealt with in a later chapter.)

Example 7

If the ratio of rise to span of a roof is to be 2:7, what would be the rise of a roof spanning 10.500 m?

The span of 10.500 m represents 7 parts of the ratio. Thus:

one part $= \dfrac{10.500}{7}$ m or 1.500 m

The rise is represented by 2 parts of the ratio. Thus:

rise   = 1.500 x 2 = 3.000 m

Example 8

A certain alloy is composed of zinc, tin and copper, in the ratio 1:6:36. What materials will be required to make 86 kg of the alloy?

total number of parts	$= 36 + 6 + 1$	$= 43$
of which one part is zinc	$= \frac{1}{43}$ of 86	$= 2$ kg
six parts are tin	$= \frac{6}{43}$ of 86	$= 12$ kg
thirty-six parts are copper	$= \frac{36}{43}$ of 86	$= 72$ kg
		total $= 86$ kg

Total $= 86$ kg of alloy: 2 kg zinc; 12 kg tin; 72 kg copper.

These examples help to illustrate the fact that fractions still have their uses, even in a metric system.

Exercise 5

1   Convert the following to percentages:
   (a) $\frac{1}{4}$
   (b) $\frac{1}{10}$
   (c) $\frac{3}{20}$
   (d) $\frac{1}{25}$
   (e) $\frac{3}{4}$

2   Convert the following to vulgar fractions:
   (a) $33\frac{1}{3}\%$
   (b) 15%
   (c) $2\frac{1}{2}\%$
   (d) 20%
   (e) 60%

3   A good slate should not absorb more than $\frac{1}{2}\%$ of its original weight of water after 12 hours soaking. What is your opinion of each of the following specimens in respect of this?
   (a)   Original weight 640 g
         Total water absorbed 60 g
   (b)   Original weight 381 g
         Total water absorbed 4 g
   (c)   Original weight 374.5 g
         Total water absorbed 2.25 g

4   What would be the total cost of 8 sheets of plywood, each 1.500 x 1.200 m at £39.20 per 10 m² plus $12\frac{1}{2}\%$ special cutting charge. This is then subject to a discount of $2\frac{1}{2}\%$ for prompt payment.

5   What is the percentage error in estimating that the total building cost of a house was £22,250 when it actually cost £22,740 to build?

6   A rectangular room measuring 3.200 m × 4.350 m is to be floored with tongued and grooved boarding costing £9.90 per m². Calculate:
   (a)   The amount of flooring you would order allowing 10% wastage (correct to first place of decimals).
   (b)   The total cost of the flooring.

7   Calculate the weights of copper and tin required to make 50 kg of bronze if the materials are to be in the ratio of 9:1 copper to tin.

8   A certain sized timber costs £1.35 per metre run. How many metres are contained in a length costing £10.50.

9   Find the value of each of the following:
   (a) $7\frac{1}{2}\%$ of £5.50
   (b) $12\frac{1}{2}\%$ of £18.48 (both to nearest $\frac{1}{2}$p)

10   After spending 15% of his pocket money an apprentice has £5.75 left. How much had he originally?

11   A flat roof in concrete 100 mm thick is 6.300 m long by 3.900 wide. What does the slab weigh if 1 m³ of the concrete weighs 2400 kg?

12   If the minimum fall of a 100 mm diameter drain in relation to its length is 1:40, what should be the total minimum fall for a drain length of 49.850 m?

13   What would be the total cost of 10,500 facings if 1500 cost £450?

14   What is the principal sum which must be invested in a builder's business at a guaranteed rate of interest of 7% per annum, in order to obtain a total income from interest of £1200 per annum?

Revision exercise 1

1   Find the prime factors of all the numbers between 17 and 28 which are multiples of 3.

2   Find the LCM and the HCF of the following:
   (a)   8, 12, 24 and 32
   (b)   4, 6, 10 and 12
   (c)   15, 45, 60 and 90

3   Convert the following to improper fractions:
   (a) $5\frac{4}{7}$ (e) $7\frac{3}{64}$
   (b) $11\frac{3}{5}$ (f) $22\frac{13}{15}$
   (c) $16\frac{7}{8}$ (g) $10\frac{27}{128}$
   (d) $8\frac{4}{9}$ (h) $3\frac{83}{144}$

4 Convert the following to mixed numbers:

(a) $\frac{171}{12}$ (e) $\frac{711}{128}$

(b) $\frac{53}{16}$ (f) $\frac{22}{7}$

(c) $\frac{235}{8}$ (g) $\frac{41}{5}$

(d) $\frac{625}{144}$ (h) $\frac{25}{4}$

5 Find the value of:

(a) $3\frac{1}{4} + 6\frac{2}{3} + 4\frac{1}{6}$

(b) $11\frac{3}{16} + 21\frac{3}{8} + 8\frac{5}{24}$

(c) $5\frac{1}{2} + 8\frac{4}{5} + 1\frac{13}{15}$

(d) $14\frac{2}{3} + 3\frac{1}{7} + 6\frac{5}{6}$

6 Find the value of:

(a) $16\frac{2}{3} - 11\frac{5}{6}$

(b) $8\frac{7}{32} - 1\frac{11}{12}$

(c) $10\frac{1}{4} - 3\frac{1}{2} - 1\frac{3}{5}$

(d) $15\frac{4}{7} - 10\frac{1}{4} - 2\frac{3}{14}$

7 Find the value of:

(a) $7\frac{1}{2} + 3\frac{2}{5} - 4\frac{11}{15}$

(b) $6\frac{1}{3} - 3\frac{1}{7} + 4\frac{1}{2}$

(c) $5\frac{11}{16} - \frac{3}{4} + 2\frac{5}{8} - 1\frac{11}{32}$

(d) $2\frac{1}{9} + 3\frac{2}{5} - 1\frac{2}{3} + 4\frac{2}{15}$

8 Evaluate:

(a) $4\frac{3}{4} \times 1\frac{1}{3}$

(b) $2\frac{13}{16} \times 5\frac{4}{15}$

(c) $9\frac{3}{5} \times \frac{15}{16}$

9 Evaluate:

(a) $8\frac{1}{15} \div 3\frac{2}{3}$

(b) $34\frac{2}{3} \div 2\frac{8}{9}$

(c) $5\frac{2}{3} \div \frac{34}{171}$

10 Simplify:

(a) 37.3 + 29.84 + 3.275 + 14.827

(b) 64 + 2.76 + 33.07 + 50.009

(c) 127.92 − 63.89

(d) 58.163 − 29.879

(e) 121.96 − 5.843 + 17.81 − 23.375

11 Find the value of:

(a) 92.62 + (11.3 − 7.81) − 76.55

(b) 12.76 + 57.31 − (11.42 + 3.14 + 8.06)

12 Evaluate, correct to two decimal places:
    (a)   17.91 × 1.207
    (b)   132.6 × 1.84
    (c)   29.81 × 0.076
    (d)   523.8 ÷ 14.84
    (e)   12.462 ÷ 6.03
    (f)   0.0792 ÷ 0.00316

13 Convert the following to decimals, correct to three places:
    (a) $\frac{13}{16}$
    (b) $\frac{3}{11}$
    (c) $\frac{7}{22}$
    (d) $\frac{14}{23}$

14 If the average thickness of a certain type of brick is 65 mm, what must be the average thickness of the bed joint so that when built the brickwork has four courses to each 300 mm?

15 The rise of each step in a staircase is 180 mm. What is the total rise (floor to floor) of the staircase if it contains fifteen steps?

16 A run of pipework contains five lengths each 790 mm long, and four bends each of which takes 140 mm. What total length of pipe is there? (Answer in metres.)

17 When soaked in water for 24 hours a brick whose dry weight was 2.720 kg was found to weigh 3.200 kg. What weight of water was absorbed per kg of brick?

18 Find the value of:
    (a)   75% of £427.67$\frac{1}{2}$
    (b)   12$\frac{1}{2}$% of 11.480 m
    (c)   15% of 50 m^3
    (d)   22$\frac{1}{2}$% of 1 tonne (= 1000 kg)

19 The ratio of cement to sand in mortar for brickwork is 1:6. What volume of sand must be used with each 50 kg of cement? (1 m^3 cement weighs 1400 kg.)

20 Lime/sand mortar for building brickwork models contains 70% sand and 30% lime. How much lime is needed to knock up 150 kg of the mortar?

21 Timber in a workshop store is estimated to contain 12% by weight of moisture. If the total weight of timber in the store is 651 kg, what part of this is water?

22 The ratio of tin to lead in a certain solder is 3:7. How many kg of each metal is contained in 35 kg of the solder?

23 The catalogue price of a tiled surround is £127.50. It is purchased by a builder who obtains 10% trade discount. Because he pays promptly he is allowed $2\frac{1}{2}\%$ cash discount on the trade price. How much does he finally pay?

24 Calculate the simple interest on £600 invested for 2 years at $4\frac{1}{2}\%$.

25 The volume of a trench for foundations is 25 m^3. When excavated the earth increases in bulk by $33\frac{1}{3}\%$. If 55% of the excavated material is required for returning to the trench after the brickwork is built, how many m^3 are there to be carted away?

5 Aids to calculation

Although a knowledge of the manipulation of numbers by the so-called long methods is an essential part of the understanding of mathematics there is no point in slogging away unnecessarily at long multiplications and divisions when we have so many short cuts to use.

Mathematical tables

A set of tables is an asset to any student and a valuable time saver. The usual contents of a set of four-figure tables includes:

1 A page of useful constants and formulae;
2 Logarithms and antilogarithms;
3 Trigonometry tables;
4 Powers, roots and reciprocals;
5 Areas of circles.

Powers and roots

A number multiplied by itself is said to be 'squared' or 'raised to the power of two'.

2 x 2 may be written as 2^2 (two squared or two to the second power). If a third factor, 2, is added we have 2 x 2 x 2 which may be written as 2^3 (two cubed or two to the third power).

In these examples 2 is the base and the small superior figure indicating the power in each case is an index (plural indices).

We use the terms *squared* and *cubed* because the most common applications of these powers are to the calculations of areas (squaring) and volumes (cubing). For this reason most mathematical tables contain these powers of numbers from 1 to 100 (usually in a combined table headed *powers, roots and reciprocals*).

They also contain tables of squares which may be used for *any* number. These give the sequence of figures required but the position of the decimal point must be determined independently. You will find such tables on pages 191-200.

Suppose we wish to evaluate 2.724^2. (Reference to a table of 'squares of numbers' is necessary in this problem.)

Look for the first two figures, 27, in the left-hand column and run your finger along that line to the column under the third figure, 2. This gives 7398. Now run your finger along further to the next set of columns headed *mean differences*. Under the fourth figure of our number, 4, we find a difference of 22 which must be added giving $7398 + 22 = 7420$.

So far we have ignored the decimal point but we now know that since the value we require is between 2^2 and 3^2 it must lie between 4 and 9, so it is in fact 7.420, that is:

$2.724^2 = 7.42$

Square roots

These are, in a sense, the reverse of powers, because:

$9 = 3^2$, 3 is the *square root* of 9.
$25 = 5^2$, 5 is the *square root* of 25.

The square root of a number is that other number which, when multiplied by itself to *two* factors gives the original number.

Care should be taken when using square root tables since numbers having the same sequence of figures do not necessarily have the same sequence of figures in their square roots.

Here is a simple example:

$$\sqrt{4} = 2$$
$$\sqrt{40} = 6.325$$
$$\sqrt{400} = 20$$
$$\sqrt{4000} = 63.25$$
$$\sqrt{40000} = 200$$

The key to this is that for each *pair* of figures in the number there is *one* figure in the square root but when there is an odd number of figures the first one is counted as a pair.

When using square root tables you will find them in two distinct parts, *square roots from 1 to 10* and *square roots from 10 to 100*. For numbers containing an odd number of figures use 'square roots from 1 to 10'. For those with an even number of figures use 'square roots from 10 to 100'. You will find such tables on pages 193–6.

Here are two examples to illustrate their use:
Consider the two square roots $\sqrt{362.7}$ and $\sqrt{3627}$.

The first has an odd number of figures (not counting the decimal part) so we need to use '1 to 10' tables. Select the first two figures in the left-hand column and run your finger along to the column under the 2, our third figure. Here we find 1903. Now go to the column headed 7 in the *mean differences* where we find 2. Add this to 1903 giving 1905. To find the position of the decimal point we apply the rule of pairing, that is, for every pair of figures in the original number there will be one figure in the square root. However, for numbers with an odd number of figures the left-hand figure will always be the odd one out and for the purpose of this rule counts as a pair.

3̲6̲2.7

Thus, we have two figures in our square root and $\sqrt{362.7} = 19.05$.
Now the second value, $\sqrt{3627}$.
Since this has an even number of figures we must use the '10 to 100' tables. Find 36 in the left-hand column and run along the line to the column under 2 to find 6017. Now go to the column under 7 in the mean differences where we find 6. Add this to 6017 giving 6023. Now do the pairing.

3̲6̲2̲7̲

There are two pairs so we have two figures before the decimal point, therefore,

$\sqrt{3627} = 60.23$.

Practice soon makes the process quick and easy.
You will have noticed that the sign for roots is $\sqrt{}$, that is, $\sqrt{9}$ is read as *the square root of 9*.
For cube root a small superior 3 is placed in front of the sign, thus $^3\sqrt{}$. For example, $^3\sqrt{27}$ is read as *the cube root of 27* – its value of course is 3 since $3^3 = 27$.

Reciprocals

A reciprocal is obtained when a number is inverted. For example, the reciprocal of 2 is $\frac{1}{2}$ (or 0.5).
To avoid tedious calculations for a value such as, say, $\frac{1}{27}$ the use of tables is called for (in this case $\frac{1}{27} = 0.03704$).
Referring to the tables on pages 191 and 192; suppose we wish to evaluate $\frac{1}{3.142}$. Find the first two figures, 31, in the left-hand column and run your finger along the line to the column under 4. This gives 3185. Now to the *mean differences* and the column under 2. This gives 2. Because our number has *increased* from 3.14 to 3.142 in finding this difference, the reciprocal will have *decreased*, so the difference must be subtracted giving 3185 − 2 = 3183.
The decimal point must be found by an analysis of the required value. The value approximated to $\frac{1}{3}$ which is 0.3, so $\frac{1}{3.142} = 0.3183$
If we had wanted $\frac{1}{31.42}$ the sequence of figures would have been the same but, since 31.42 is ten times 3.142 its reciprocal is one tenth, that is, $\frac{1}{31.42} = 0.03183$.

Exercise 6

1 Use tables of squares to find the value of:
(a) $(28)^2$
(b) $(3.27)^2$
(c) $(42.13)^2$
(d) $(185)^2$
(e) $(0.85)^2$
(f) $(12.25)^2$

2 Find from the tables the square roots of the following numbers:
(a) 173
(b) 13.5
(c) 2401
(d) 0.275
(e) 12.25
(f) 841

3 Evaluate the following from the reciprocal tables:
(a) $\frac{1}{23}$
(b) $\frac{1}{129}$
(c) $\frac{1}{0.81}$
(d) $\frac{1}{5840}$
(e) $\frac{1}{0.36}$
(f) $\frac{1}{105}$

4 Find the value of each of the following expressions:
(a) $(3.751)^2 + (67.2)^2 - \sqrt{2030}$
(b) $\frac{1}{25} + \frac{1}{18} - (0.52)^2 + \sqrt{17.8}$
(c) $(28.12)^2 - \sqrt{36.24} + \frac{1}{0.435}$

Logarithms

Returning to the idea of indices and powers let us suppose that we wish to find the value of $7^3 \times 7^2$:

$$7^3 = 7 \times 7 \times 7$$
$$\text{and } 7^2 = 7 \times 7$$
$$\text{so } 7^3 \times 7^2 = 7 \times 7 \times 7 \times 7 \times 7$$
$$= 7^5$$

This result is 7 to the power of 5. Can you see the connection between the original and the result in index form? We could have arrived at the solution simply by *adding* the indices. That is:

$$7^3 \times 7^2 = 7^{3+2}$$
$$= 7^5$$

This will always apply providing the base number is the same. In this case the base number was 7.

Here is another example:

$$8^4 \times 8^2 \times 8^3 = 8^{4+2+3}$$
$$= 8^9$$

The result is 8 to the power of 9. This is only the first step in changing a multiplication into an addition. At this stage the answer remains as a power of the base number.

Division by indices

Let us now see how we can apply indices to a division problem. Suppose we wish to find the value of $6^4 \div 6^2$

$$6^4 \div 6^2 = \frac{6^4}{6^2} = \frac{6 \times 6 \times \cancel{6} \times \cancel{6}}{\cancel{6} \times \cancel{6}}$$
$$= 6 \times 6 \text{ (by cancelling)}$$
$$= 6^2$$

Again look for the connection between the original and the result and you will see that this time you arrived at the same answer by simply subtracting the indices. That is:

$$6^4 \div 6^2 = 6^{4-2}$$
$$= 6^2$$

Again, this will apply for any similar problem provided the base number is the same. In this case the base number was 6.

Here is another example:

$$12^5 \div 12^2 = 12^{5-2}$$
$$= 12^3$$

Now a composite example:

$$\frac{9^3 \times 9^2 \times 9^5}{9^4 \times 9^4} = \frac{9^{3+2+5}}{9^{4+4}}$$
$$= \frac{9^{10}}{9^8}$$
$$= 9^{10-8}$$
$$= 9^2$$

From the example above, we see that if we can express numbers in index form to a common base, we are able to replace multiplication by addition, and division by subtraction.

We have in fact establisned the first two 'laws of indices':
1 To multiply *add* the indices
2 To divide *subtract* the indices

Logarithms are indices and the tables of logarithms on pages 197-8 will enable you to express any number in index form to the base 10. Logarithms to the base 10 are called *common logarithms*. The law of indices already explained, also apply to logarithms.

Finding the logarithm of a number

A logarithm is always in two parts: a whole number part (in front of the decimal point) called the *characteristic*; and a fractional part (after the decimal point) called the *mantissa*. For example:

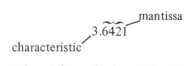

Only the mantissa is found from the log tables. The characteristic, which depends on the size of the number, you must find yourself (this will be explained later).

First, to find the mantissa from the tables proceed as follows. Read the first two figures of the number in the column on the extreme left of the table. Run your finger along that line to the column which is headed with the third figure of the number. Here you will find a four figure number. Add to this the number in the *mean difference* column under the fourth figure of your number and the result is the required mantissa.

Example 1

Find the mantissa of log 5392.
To save continually turning the pages an extract from the log tables is printed on page 50.

Looking along the line from 53 and under 9 we find 7316. In the same line in the *mean difference* column under 2 we find a 2 which, added to 7316, gives 7318. That is:

the mantissa of log 5392 is .7318

As an exercise follow the same procedure to find log 5245 which you will find to be .7197 (that is, its mantissa).

These tables cater for numbers containing up to four significant figures only. Any number containing more than this must be corrected to four significant figures before its log is found. If there are less than four figures there is no problem. For example the mantissa of the logarithm of 50 is found in the *0* column in the *50* line. The mean difference columns are not used in this case.

Extract from log tables

	0	1	2	3	4	5	6	7	8	9	123456789
50	.6990	6998	7007	7016	7024	7033	7042	7050	7059	7067	123345678
51	.7076	7084	7093	7101	7110	7118	7126	7135	7143	7152	123345678
52	.7160	7168	7177	7185	7193	7202	7210	7218	7226	7235	122345677
53	.7243	7251	7259	7267	7275	7284	7292	7300	7308	7316	122345667
54	.7324	7332	7340	7348	7356	7364	7372	7380	7388	7396	122345667
55	.7404	7412	7419	7427	7435	7443	7451	7459	7466	7474	122345667

Now the characteristic must be found. Its value is one less than the number of figures in the whole-number part of the number.

The characteristic of logarithms:

1   1.75   = 0.
2   61.27 = 1.
3   569.2   = 2.

We are now able to find the complete log of a number as follows:

log 5392 = 3.7318

3 being the characteristic since the number has four figures and .7318 being the mantissa from the tables as previously explained.

You should now make yourself familiar with the tables by jotting down a string of numbers and then looking up their logarithms.

Exercise 7

1   Write down the answers to the following in index form:

(a) $15^3 \times 15 \times 15^2$

(b) $\dfrac{x^2 \times x^4}{x^3}$

(c) $\dfrac{8^{3.2} \times 8^{1.6}}{8^{1.8}}$

(d) $\dfrac{16^{\frac{3}{4}} \times 16^{1\frac{1}{4}}}{16}$

2   Write down the characteristics only of the following numbers:
   (a)   73.62
   (b)   7362
   (c)   5.94
   (d)   100
   (e)   10000

3   Find the logarithms of each of the following numbers:
  (a)   25   (h)   27
  (b)   34.16   (i)   8.66
  (c)   2.974   (j)   1.414
  (d)   1.732   (k)   2.54
  (e)   3.142   (l)   9658
  (f)   1728   (m)   25991
  (g)   6.25   (n)   76879

4   Compare the logarithms of: 3.333; 33.33; 333.3; 3333. What do you particularly notice about them?

Calculation by logarithms

The following examples illustrate a widely accepted method of setting out logarithms for calculations. A table of two columns is used, one headed *No.* and the other *Log.* It is neat, tidy and easy to use and check.

Example 2

Find the value of 29.32 x 1.954 x 16.72.
   To multiply these together their logs must be added. Write the numbers one under the other in the *log* column. Against each, in the *log* column write its logarithm. Add the logs and this gives the log of the required product, that is 2.9812. This must now be converted back to a number, the *reverse process* of finding a logarithm. It could be done by using the log tables in reverse but as this is rather a tedious procedure the tables are printed in reverse and called *antilogarithms.* Therefore, reading the antilogs forwards is equivalent to reading the log tables in reverse. See antilog tables on pages 199–200.

No.	Log
29.32	1.4672
1.954	0.2908
16.72	1.2232
957.6	2.9812

   Now, remembering that only the mantissa is found from the tables, read the four figures in this mantissa (.9821) in the antilogs. Find 98 in the left-hand column, run your finger along that line to the *1* column and you find 9572. Now in the *mean difference* column under 2 you find the figure 4 which is added to the 9572 giving 9576. The characteristic is 2, which tells us that there are three figures in the number.

If the characteristic is one less than the number of figures in the number, it follows that the number of figures in the number will be one more than the value of the characteristic.

The answer is therefore 957.6.

Example 3

Evaluate $\dfrac{3.924 \times 1.05 \times 27.2}{24.6 \times 1.52}$

It is a good idea to carry out a rough check mentally using the nearest *convenient* whole numbers. This example could be expressed very approximately as:

$$\frac{4 \times 1 \times 27}{24 \times 1\frac{1}{2}} = \frac{108}{36}$$
$$= 3$$

This gives the rough size of the answer and leaves no doubt about the position of the decimal point. Now for the actual problem. First find the logs of the three numbers in the numerator and add them together. This gives the log of the whole of the numerator. We do not need to find the antilog of this. Next find the logs of the two numbers in the denominator and add them together. The result is the log of the whole denominator. Now subtract the log of the denominator from that of the numerator thereby carrying out the division. Now antilog and position the decimal point.

The answer is 2.997.

No.	Log
3.924	0.5937
1.05	0.0212
27.2	1.4346
num.	2.0495
24.6	1.3909
1.52	0.1818
denom.	1.5727
num.	2.0495
denom.	1.5727
2.997	0.4768

Powers and roots by logarithms

As an alternative to the use of tables, *powers* and *roots* may be found using logarithms.

Example 4

Consider $(25.4)^2$.

If we write it as 25.4 × 25.4 we see that the evaluation could be carried out by adding log 25.4 to log 25.4 and then antilogging. This is quite acceptable but it is easier to multiply log 25.4 by 2 instead.

No.	Log	
25.4	1.4048	
	2	(multiply by 2)
645.1	2.8096	

Therefore, the answer is 645.1.

Example 5

To find the value of 8.76^3, multiply the log by 3.

No.	Log	
8.76	0.9425	
	3	(multiply by 3)
672.2	2.8275	

The answer is 672.2.

Since we had to multiply the log of a number by 2 to square it, it follows that we must divide the log by 2 when evaluating a square root. Similarly we divide the log by 3 when finding a cube root. The following examples illustrate this application.

Example 6

To find $\sqrt{259.5}$, divide the log by 2.

No.	Log	
259.5	2.4142	
	2)2.4142	(divide by 2)
16.11	1.2071	

The answer is 16.11.

Example 7

Find the value of $\sqrt{237.5}$ by dividing the log by 3.

No.	Log	
237.5	2.3756	
	3)2.3756	(divide by 3)
6.193	0.7919	

The answer is 6.193.

Example 8

Evaluate the expression $\dfrac{(15.75)^2 \times \sqrt{8.65}}{(2.316)^3}$

No.	Log	
15.75	1.1973	
	2	(multiply by 2)
15.75^2	2.3946	
8.65	2)0.9370	(divide by 2)
$\sqrt{8.65}$	0.4685	
2.316	0.3647	
	3	(multiply by 3)
2.316^3	1.0941	
15.75^2	2.3946	
$\sqrt{8.65}$	0.4685	
	2.8631	
2.316^3	1.0941	
58.75	1.7690	

Find the powers and roots first. This is carried out by first multiplying the log by 2 to find 15.75^2. Then dividing the log by 2 to find the square root of 8.65. Finally, multiplying the log by 3 to find 2.316^3. With the powers and roots solved, the expression can now be evaluated by multiplying and dividing.

Therefore, the value of expression is 58.75.

Exercise 8

Use logarithms to evaluate the following questions.

1   (a)   51.26 x 16.3

(b)   53.23 x 5.545

(c)   274 x 1.542

2   (a) $\dfrac{16.19}{4.26}$

(b) $\dfrac{3657}{96}$

(c) $\dfrac{3.975}{1.842}$

(d) $\dfrac{879500}{4657}$

3   (a) $\dfrac{15.76 \times 324.8}{263.7}$

(b) $\dfrac{195 \times 76.84}{2438}$

(c) $\dfrac{357.1}{9.684 \times 7.39}$

4   (a) $(6.375)^2$

(b) $(45.5)^2$

(c) $(2.38)^3$

(d) $(1.625)^4$

5   (a) $\sqrt{4608}$

(b) $\sqrt{36400}$

(c) $\sqrt{53.27}$

(d) $\sqrt{1.85}$

6   (a) $\dfrac{(1.75)^2 \times \sqrt{6.452}}{1.875}$

(b) $\dfrac{(2.945)^3 \times (1.375)^2}{\sqrt{218.5}}$

(c) $\dfrac{\sqrt{75.9} \times \sqrt{95.7}}{25.8}$

(d) $\dfrac{365 \times (40.3)^2}{(3.65)^3 \times \sqrt{6875}}$

Logarithms of numbers less than 1

So far we have used logarithms of numbers greater than 1 but it is often necessary to deal in fractions. The difference between the logarithms of whole numbers and fractions lies only in the characteristic. This is the part of the logarithm before the decimal point. For numbers greater than 1 the characteristic is a positive number but for numbers less than 1 it is negative. However, the mantissa of the logarithm (the part obtained from the tables) is *always* positive. Because of this the negative characteristic is written with its minus sign above it rather than in front of it. To obtain the characteristic we add 1 to the number of noughts following the decimal point before the figures start.

Consider the decimal fraction 0.00765.

This has two noughts after the decimal point and therefore the characteristic of its log is – 3 but it is written as $\bar{3}$, and read as 'bar three'.

The mantissa is obtained by reading the 765 into the tables in the usual way giving 8837, so we have:

log 0.00765 = $\bar{3}$.8837

Because the characteristic is negative while the mantissa is positive, great care is needed when using the logarithms for evaluations.

Use your log tables to check the following:

1 log 0.00725 = $\bar{3}$.8603 (two noughts, $\bar{3}$)
2 log 0.08216 = $\bar{2}$.9146 (one nought, $\bar{2}$)
3 log 0.7901 = $\bar{1}$.8977 (no nought, $\bar{1}$)

Here are some worked examples illustrating the use of negative characteristics.

Example 9

Evaluate $\dfrac{0.75 \times 1.16}{0.925}$.

No.	Log
0.75	$\bar{1}$.8751
1.16	0.0645
	$\bar{1}$.9396
0.925	$\bar{1}$.9661
0.9408	$\bar{1}$.9735

The answer is 0.9408.

Example 10

Find the value of $\dfrac{3.75\pi}{0.875}$.

No.	Log
3.75	0.5740
π	0.4972
	1.0712
0.875	1.9420
13.47	1.1292

The answer is 13.47.

Note: $\bar{1} + 1 = 0$; $0 - \bar{1} = 1$ (the double negative creates a positive).

Example 11

Evaluate $\dfrac{(0.83)^2}{\sqrt{4.8}}$.

The answer is 0.3145.

No.	Log
0.83	$\bar{1}$.9191
	2
$(0.83)^2$	$\bar{1}$.8382
$\sqrt{4.8}$	0.3406
0.3145	$\bar{1}$.4976
4.8	0.6812
÷ 2	0.3406

(multiply by 2)

(divide by 2)

Note: When multiplying $\bar{1}$.9191 by 2, the $2 \times \bar{1} = \bar{2}$ and the 1 to carry (from 2×0.9) makes it $\bar{1}$ since $\bar{2} + 1 = \bar{1}$.

Example 12

Find the square root of 0.9736.

This example illustrates two important points. The first is that the square root of a number less than one is always greater than the number itself. The second is that a logarithm with a negative characteristic must be treated as two separate values, the negative characteristic, and the positive mantissa. They can never be read as one value.

No.	Log
0.9736	$\bar{1}$.9884
	$\bar{1}$
	2.9884
	$\bar{2}$ + 1.9884
	$\bar{2}$
	1.9884
0.9868	$\bar{1}$ .9942

(subtract 1)

(add 1)

(divide by 2)

When dividing log 0.9736 by 2 you first make the characteristic the next lowest number exactly divisible by 2, in this case $\bar{2}$. In so doing you have *subtracted* 1 from the characteristic. To retain the overall value of the log you must add that 1 back to the mantissa which becomes 1.9884. We now have $\bar{2}$ and 1.9884 which together are the same as $\bar{1}$ and 0.9884.

The two parts, negative and positive can now be separately divided by 2. Thus:

$$\sqrt{0.9736} = 0.9868$$

Exercise 9

1   Write down the logarithms of the following from your log tables:
   (a)   0.07038
   (b)   0.1682
   (c)   0.00954
   (d)   0.0909
   (e)   0.000012
   (f)   0.043

2   Find the numbers whose logs are:
   (a) $\bar{1}.8714$
   (b) $\bar{2}.9360$
   (c) $\bar{5}.7019$
   (d) $\bar{3}.3434$
   (e) $\bar{1}.1726$
   (f)   0.7621

3   Evaluate:
   (a)   5.278 x 0.0962
   (b)   0.851 x 1.414 x 0.975
   (c)   27.8 x 0.0087 x π
   (d)   0.667 x 0.333 x 122.5

4   Evaluate:
   (a)   2.035 ÷ 72.9
   (b)   0.3781 ÷ 1.732
   (c)   1.875 ÷ 0.7765
   (d) $\dfrac{1}{15.86}$
   (e) $\dfrac{10}{273.3}$

5   Use logs to find the value of:
   (a) $\dfrac{18.3 \times 0.846 \times 1.25}{0.98 \times 213.5}$
   (b) $\dfrac{0.9327 \times 0.6358 \times 0.0711}{0.084 \times 0.0074}$

6   Find the following roots by logs:
   (a) $\sqrt{0.7321}$
   (b) $\sqrt{0.0186}$
   (c) $\sqrt[3]{0.0979}$
   (d) $\sqrt[4]{0.2461}$
   (e) $\sqrt[3]{0.00864}$
   (f) $\sqrt{0.999}$

7   Evaluate correct to two places of decimals:
   (a) $\sqrt[3]{\dfrac{0.6301 \times 6.359}{9.817 \ \times 0.793}}$

   (b) $\dfrac{8.31 \times (4.82)^2 \times (0.076)^3}{\sqrt{0.88} \times 17.81}$

   (c) $\dfrac{(0.6125)^{2/5} \times (1.119)^{1/2}}{0.81 \ (17.54 + 3.68)}$

Electronic calculators

Now that electronic calculators are readily available at relatively low cost it is reasonable that anyone with the need to make calculations will use one. Earlier prejudices against their use seem to have been dispelled but it is important that you do not become a slave to a calculator, and allow your mind to become sluggish. Simple operations can be performed more quickly in the head!

The simplest models have only the basic arithmetical processes. Others give squares, square roots, reciprocals and percentages. At the top of the range many extremely complicated processes can be operated. In the following examples it is assumed that the student has a basic model with only + − × ÷ and % functions. The sequence of operations is given followed by the display obtained at each operation and finally the result when the = key is used.

Example 1

Find 23.54 + 17.62.

   Sequence of keys: 2 3 . 5 4 + 1 7 . 6 2 =

Key	*Display*
2	2.
3	23.
.	23.
5	23.5
4	23.54

Key	Display
+	23.54
1	1.
7	17.
.	17.
6	17.6
2	17.62
=	41.16

Example 2

Find 17 + 8.5 – 3.6.

Sequence of keys: 1 7 + 8 . 5 – 3 . 6 =

Key	Display
1	1.
7	17.
+	17.
8	8.
.	8.
5	8.5
–	25.5
3	3.
.	3.
6	3.6
=	21.9

Example 3

Find 13.5 x 2.4.

Sequence of keys: 1 3 . 5 x 2 . 4 =

Key	Display
1	1.
3	13.
.	13.
5	13.5
x	13.5
2	2.
.	2.
4	2.4
=	32.4

Example 4

Find $\frac{19 \times 7.8}{3.5}$.

Sequence of keys: 1 9 x 7 . 8 ÷ 3 . 5 =

Key	Display
1	1.
9	19.
x	19.
7	7.
.	7.
8	7.8.
÷	148.2
3	3.
.	3.
5	3.5
=	42.342857 (42.34 to two decimal places)

Example 5

Find $\frac{1.75 \times 12.3}{1.16 \times 6.4}$.

Sequence of keys: 1 . 7 5 x 1 2 . 3 ÷ 1 . 1 6 ÷ 6 . 4 =

Key	Display
1	1.
.	1.
7	1.7
5	1.75
x	1.75
1	1.
2	12.
.	12.
3	12.3
÷	21.525
1	1.
.	1.
1	1.1
6	1.16
÷	18.556034
6	6.
.	6.
4	6.4
=	2.8993803 (2.90 to two decimal places)

Example 6

Find $(15.61)^2$.

Sequence of keys: 1 5 . 6 1 × 1 5 . 6 1 =

Key	Display
1	1.
5	15.
.	15.
6	15.6
1	15.61
×	15.61
=	243.6721

(Note that the number 15.61 is not repeated, the multiplication key followed by the equals key produces the square.)

By the same procedure, as shown in the above example, any power may be found. Feed in the number followed by the x key and then press the = key one time less than the index of the required power. That is, for the fourth power the = key would be pressed three times after the x key.

Example 7

Find $(3.14)^3$.

Sequence of keys: 3 . 1 4 × = =

Key	Display
3	3.
.	3.
1	3.1
4	3.14
×	3.14
=	9.8596
=	30.959144 (30.95 to two decimal places)

Example 8

Find $\dfrac{1}{2.5}$.

Sequence of keys: 1 ÷ 2 . 5 =

The answer is 0.4.

If your calculator has a reciprocal key (marked $\frac{1}{x}$) the process is simpler. Just enter 2.5 followed by the reciprocal key.

Example 9

Find the value of 18% of £325. (Remember *of* means *multiply*.)

    Sequence of keys: 1 8 % x 3 2 5 =

The answer is £58.50.

Example 10

Add a surcharge of 15% (VAT) to £178.50.

    Sequence of keys: 1 7 8 . 5 + 1 5 % =

Key	Display
1	1.
7	17.
8	178.
.	178.
5	178.5
+	178.5
1	1.
5	15.
%	26.775
=	205.275 (£205.27½)

Example 11

Evaluate £123 + £46 + £1.50 less 12.5% discount.
    Enter the three amounts and follow with the sequence:

– 1 2 . 5 % =

    When the – key is pressed the display will give the sum before discount. The result after discount is £149.1875 (£149.19 to the nearest penny).
    Every calculator has an instruction booklet supplied with it. Study this carefully to obtain the best results from your calculator. No exercises are given with this section but you can, of course, usefully employ your calculator on any of the exercises in this book.

6 Perimeters and areas of plane figures

Many common problems associated with the building trades involve the calculation of *perimeters, areas* and *volumes*. In this chapter we shall deal with the perimeters and areas of the more common shapes which you are likely to meet.

The *perimeter* of a figure is the distance all the way round, or the length of its boundary (measured in metres and millimetres); it is a length or *linear* measurement.

The *area* of a figure is found by multiplying two linear (length) measurements to give an answer in square measure, that is, square metres or square millimetres. This is also known as 'superficial' measurement and the term *metres super* may be used.

The *square*, then, is the unit of area. It has four equal sides and four right angles.

The area of a square = *length of side* × *length of side*

If we denote length of side by the letter *l* and area by the letter *A* we can write:

$A = l^2$

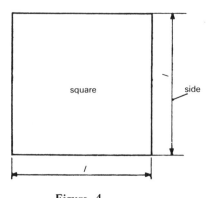

Figure 4

The *rectangle* is a common shape in building work.

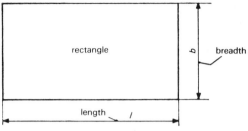

Figure 5

The *area* of a rectangle = *length* × *breadth*

or, $A = l \times b$

A large number of problems may be based on rectangular shapes. Study the following examples carefully.

Example 1

Calculate the total area and perimeter of the shape in Figure 6.

To find the area, divide the figure into a convenient number of rectangles, calculating the area of each. (There are usually several different methods of doing this.)

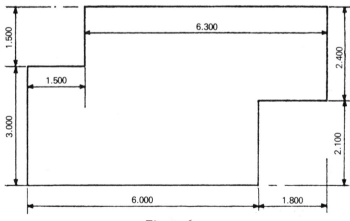

Figure 6

Method 1

You could say:

Area required = A + B + C, or

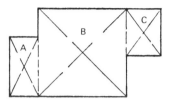

Figure 7

Method 2
Area required = total area of rectangle - (D + E)

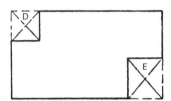

Figure 8

Using method 2:

Total area of rectangle $= 7.800 \times 4.500 = 35.10 \text{ m}^2$
$\qquad$ D + E $= (1.500 \times 1.500) + (1.800 \times 2.100)$
$\qquad\qquad = 2.25 + 3.78 = 6.03 \text{ m}^2$
$\qquad$ Area required $= 35.10 - 6.03 = 29.07 \text{ m}^2$

Figures 9 and 10 show a useful method of calculating the perimeter of this and similar examples.

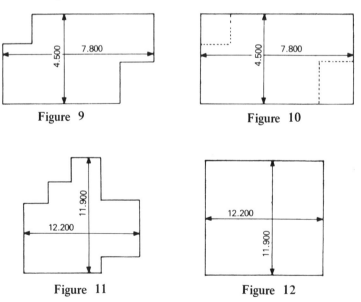

Figure 9

Figure 10

Figure 11

Figure 12

Do you see that Figures 9 and 10 have perimeters of identical length?

Similarly, the perimeters are the same in Figures 11 and 12. Applying this in the first example (Figure 6):

perimeter = 2 (maximum length) + 2 (maximum breadth)
= (2 x 7.800) + (2 x 4.500)
= 15.600 + 9.000
= 24.600 m

Thus in all cases where there are *rectangular corner insets* to the figure, the total perimeter will be equal to that of *a rectangle* with the maximum dimensions of the figure.

Note: This is not the case where any inset in the length or breadth is *not* at a corner.

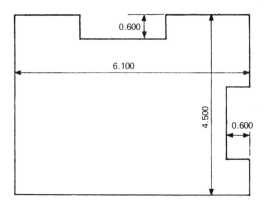

Figure 13

In Figure 13 the perimeter is equal to that of an ordinary rectangle of dimensions 6.100 m x 4.500 m *plus* the *depth of each inset line.* That is:

perimeter = (2 x 6.100) + (2 x 4.500) + (4 x 0.600)
= 12.200 + 9.000 + 2.400
= 23.600 m

Example 2

Calculate the total area and perimeter of the shape in Figure 14. For the area, the rectangle can be completed as shown in Figure 15.

Area A = 1.500 x 1.800 = 2.700 m^2
Area B = 2.100 x 0.750 = 1.575 m^2
Area C = 3.300 x 0.900 = 2.970 m^2

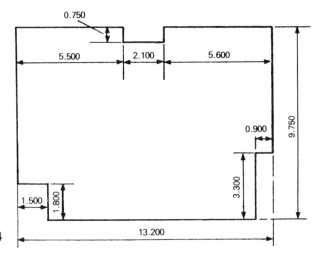

Figure 14

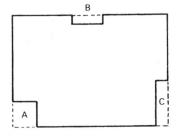

Figure 15

Area required   = total area of rectangle – (A + B + C)
                  = (13.2 x 9.750) – (2.7 + 1.575 + 2.97) m²
                  = 128.7 –7245 m²
                  = 121.455 m²

Total perimeter   = 2(13.2 + 9.75) + (2 x 0.75) m
                  = 45.9 + 1.5 m
                  = 47.400 m

Area of a triangle

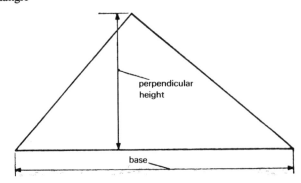

Figure 16

Area of any triangle $= \dfrac{base \times perpendicular\ height}{2}$

Note carefully that it is the *perpendicular height* which is used in the formula. Mistakes can occur, particularly when the triangle is of the type shown in Figure 17.

Area $= \dfrac{70 \times 40}{2}$ mm²

$= \dfrac{2800}{2}$

$= 1400$ mm²

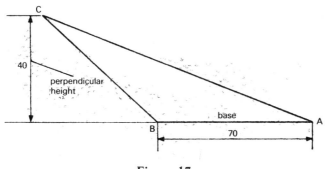

Figure 17

An alternative method of finding the area of a triangle when the three sides are given is known as the *sum formula of the area of a triangle* (sometimes called the *S formula*).

You will see from Figure 18 that the letters a, b and c represent the lengths of the sides of the triangle.

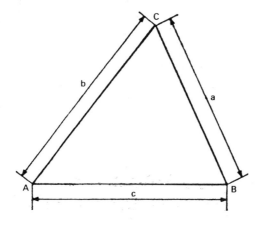

Figure 18

$$s = \frac{a + b + c}{2}$$, the semi-perimeter, and the area of the triangle

$$= \sqrt{s\,(s - a)\,(s - b)\,(s - c)} \text{ square units}$$

Right-angled triangles

The triangle is a most important shape in building work, mainly because it is the only figure which will keep its shape (unless there is a distortion or change in length of one of its members).

This can be illustrated in many ways.

Sketch six examples of triangulation applied to building construction. Use the following list to help you with ideas:

1 Roof construction
2 Brackets
3 Bracing to frames and strutting to floor joists
4 Scaffolding

A right-angled triangle is any triangle which contains a right angle (an angle of 90°); a right-angled triangle can easily be constructed from an acute angle by drawing a line perpendicular from the base to meet the other line enclosing the angle.

Note the following points connected with the right-angled triangle.

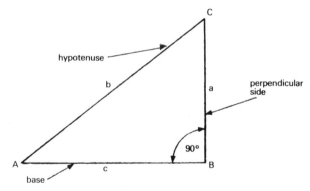

Figure 19

1 Letter the vertices (angles) with the *capital letters* of the alphabet, and the opposite sides with the corresponding small letter.

Thus, in Figure 19 we have a right-angled triangle ABC, with sides of a, b and c units respectively.

2 Notice in Figure 19 the names given to the respective sides. The *hypotenuse* is always the side facing the right angle.

3 In any triangle, the internal angles add up to 180°. In a right-angled triangle, of course, the other two angles must add up to 90°.

A right-angled triangle of particular importance in building calculations is the well-known 3:4:5 triangle.

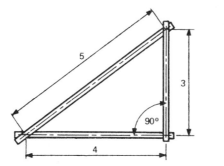

Figure 20

Any triangle whose sides are in the ratio 3:4:5 is, in fact, a right-angled triangle, the right angle being the angle opposite to the longest (or 5) side, which is, of course, the hypotenuse.

This is very useful and can be employed very often in setting out a right angle. Figure 20 shows a typical frame constructed for this purpose.

A very interesting point concerning the 3:4:5 triangle is illustrated in Figure 21.

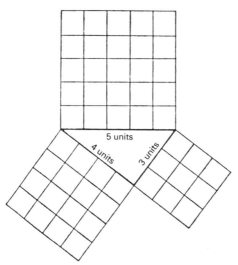

Figure 21

If a square is constructed on each side and divided into small unit squares, the square on the hypotenuse contains 25 units and the other two contain 9 units and 16 units respectively, which add up to 25.

Another way of expressing this is:

$$3^2 + 4^2 = 5^2$$

It is in fact true of every right-angled triangle that *the square on the hypotenuse is equal to the sum of the squares on the other two sides.*

This is a very important statement and is known as the theorem of Pythagoras. Pythagoras was an Ancient Greek mathematician who is thought to have first discovered this fact.

Note: All right-angled triangles are not 3:4.5 triangles.

We can use the theorem of Pythagoras to find the third side af any right-angled triangle if we know the other two sides.

Example 3

A ladder stands against a vertical wall on level ground with its foot 1.5 m from the wall and its top at a height of 3.6 m. Find the length of the ladder.

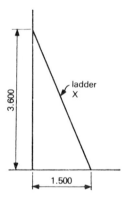

Figure 22

Figure 22 illustrates the problem. Call the length of the ladder x.

Then x^2 = $1.5^2 + 3.6^2$
Then x^2 = $1.5^2 + 3.6^2$
= 2.25 + 12.96
= 15.21

Now, this is like finding the side of a square whose area is 15.21 square units (see page 64) and x will be the square root of 15.21.

$x = \sqrt{15.21}$
= 3.9 m

Example 4

A guy wire supporting a flag mast 9 m high has its lower end anchored at a point 2 m from the foot of the mast. How long is the wire?

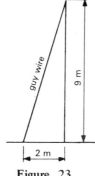

Figure 23

Referring to Figure 23, and calling the length of the wire x m,

$$x^2 = 2^2 + 9^2$$
$$= 4 + 81$$
$$= 85$$
$$x = \sqrt{85}$$
$$= 9.219 \text{ m}$$

Example 5

A triangular building plot with a right angle at one corner has its shortest side 18 m long and its longest side 30 m long. What is the length of its third side?

You will notice that we are this time *given* the hypotenuse and have to find one of the other sides.

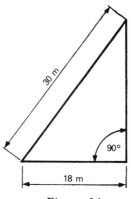

Figure 24

Call the unknown side x m. Then, by Pythagoras:

$$x^2 + 18^2 = 30^2$$

It follows that:

$$x^2 = 30^2 - 18^2$$
$$= 900 - 324$$
$$= 576$$
$$x = \sqrt{576}$$
$$= 24 \text{ m}$$

Example 6

The hypotenuse of a right-angled triangle is 15 cm and one of the other sides is 12 cm. What is the length of the third side?

Calling the unknown side x cm,

$$x^2 + 12^2 = 15^2$$
$$x^2 = 15^2 - 12^2$$
$$= 225 - 144$$
$$= 81$$
$$x = \sqrt{81}$$
$$= 9 \text{ cm}$$

Note: Do you recognize this as a 3:4:5 triangle? It is, because the sides are 3 x 3, 3 x 4 and 3 x 5, which are, therefore, in the ratio 3:4:5.

Area of a parallelogram

A parallelogram is a four-sided figure having its *opposite sides parallel*. For example:

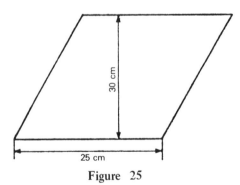

30 cm

25 cm

Figure 25

Area = *length of base side* x *perpendicular height*
$$= 25 \times 30 \text{ mm}^2$$
$$= 750 \text{ mm}^2$$

Area of a trapezium

A trapezium is a four-sided figure with two *sides parallel.* For example:

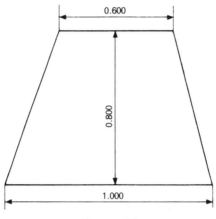

Figure 26

Area = *half the sum of the parallel sides × the perpendicular distance between them*

$$= \frac{1.000 + 0.600}{2} \times 0.800$$

$$= 0.800 \times 0.800$$

$$= 0.640 \text{ m}^2$$

In all the previous examples, the perimeter would be found by simple addition, knowing the length of the sides.

Composite figures

You now have sufficient information to solve a wide variety of problems involving a combination of the previous shapes. Regular and irregular polygons (figures with five or more sides) can usually be divided into convenient areas for calculation in the following way.

Example 7

Calculate:
(a)   The total area in the plot shown in Figure 27;
(b)   The total perimeter.

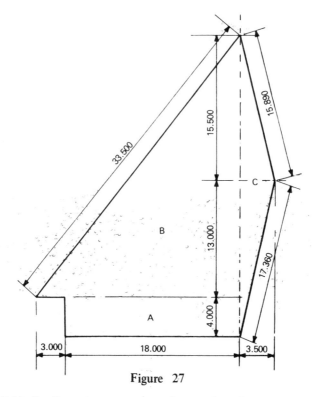

Figure 27

First, divide the figure into a number of convenient shapes.

Total area = A + B + C

$$= (18 \times 4) + \frac{21 \times 28.5}{2} + \frac{32.5 \times 3.5}{2}$$

$$= 72 + 299.25 + 56.875$$

$$= 428.125 \text{ m}^2$$

Total perimeter = 18 + 17.36 + 15.89 + 33.5 + 3 + 4

= 91.750 m

The circle

At this stage we shall deal only with the method of finding the *area* and *perimeter* of the circle and its application to practical problems.

Note: The *perimeter* of a circle has a special name; it is called the *circumference*.

The distance from the centre of a circle to any point on the circumference is called the *radius*.

The *diameter* of a circle is a line drawn through the centre to meet the circumference at both ends.

Obviously, then: *diameter = twice radius* (see Figure 28).

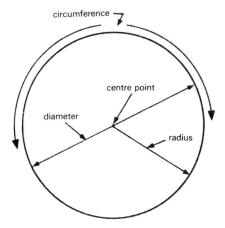

Figure 28

Circumference of a circle

The circumference of a circle has a direct relationship to its radius or diameter.

Draw a circle of any radius, mark off the diameter round the circumference, and you will find that for a circle of any size, *length of circumference = approximately $3\frac{1}{7}$ times the diameter*.

The number of times that the diameter will go into the circumference is actually 3.14159 (to six significant figures), and this amount is usually expressed by the Greek leter π (pi), pronounced *pie*.

For most normal purposes the value of π may be taken as 3.14 if using decimals, or $3\frac{1}{7}$ if using fractions.

From the foregoing it should be obvious that:

circumference of a circle $= \pi \times diameter$
or $2\pi \times radius$

Area of a circle

Here is a simple method of deriving a formula for the area of a circle.

Draw a circle of any radius on a piece of scrap paper and divide the circumference into a fairly large number of equal parts. (The greater the number of parts taken, the more accurate the final result will be.)

Now draw radial lines back to the centre from each point on the circumference (see Figure 29).

Cut along the line AB (making two semi-circles) and then cut up each radial line almost to the circumference. Now open out each semi-circle and you will find that one half will fit into the other as shown in Figures 30 and 31.

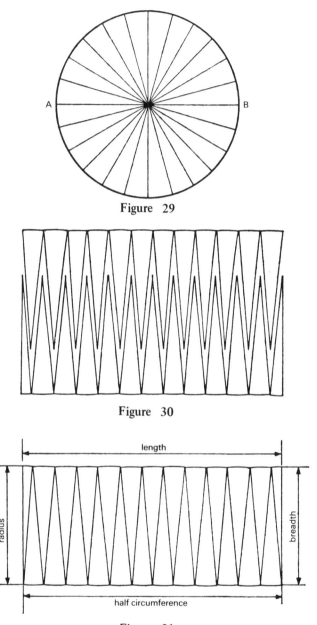

Figure 29

Figure 30

Figure 31

From Figure 31 you will see that the circle has become approximately a rectangle whose length is half the circle's circumference and whose breadth is the circle's radius.

Since the area of a rectangle = length x breadth:

Area of a circle $= \dfrac{2\pi \times \text{radius} \times \text{radius}}{2}$

$= \pi \times \text{radius} \times \text{radius}$

$= \pi \text{ radius squared}$

This is the exact formula for the area of any circle, thus: if we call the radius R and the diameter D, then:

Circumference of a circle $= 2\pi R$ or πD

Area of a circle $= \pi R^2$ or $\dfrac{\pi D^2}{4}$

Example 8

A large semi-circular bay window is built as an addition to a room. The diameter of the bay is 3.200 m. Calculate:

(a)   The floor area;
(b)   The length of curved skirting.

Here we need *half* the area of a circle and *half* its circumference. The area of a circle is given by:

$A = \pi R^2$ and R is 1.600 m

so, for the semi-circle:

$A = \tfrac{1}{2}\pi R^2$

$= 0.5 \times 3.142 \times 1.6 \times 1.6 \text{ m}^2$
$= 4.022 \text{ m}^2$

Circumference of circle $= 2\pi R$
Length of skirting $= \pi R$ (semi-circle)
$= 3.142 \times 1.6 \text{ m}$
$= 5.027 \text{ m}$

Exercise 10

1   Find by a quick method the total perimeter of the shape shown in Figure 32. Check your answer by another method.

2   The area shown in Figure 32 is to be paved with 150 mm square floor tiles. Estimate the number required to do the job, allowing $2\tfrac{1}{2}\%$ for wastage.

3   What is the length of the intrados of a semi-circular brick arch spanning 750 mm?

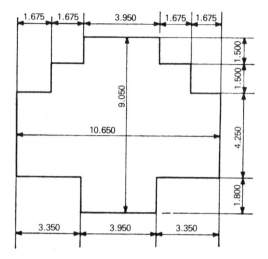

Figure 32

4   Determine the total area of sheet lead required to make 145 lead flashing pieces, each piece shown as in Figure 33.

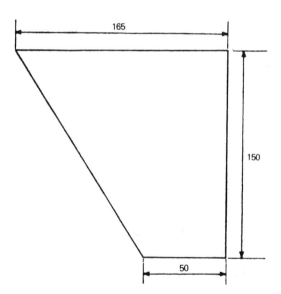

Figure 33

5   Find the total area in square metres of the wall shown in Figure 34.

6   If sheet zinc costs £13.50 per m², what is the cost of a rectangular piece measuring 1.250 m x 2.750 m?

7   A plot of land is in the shape of a right-angled triangle; the sides containing the right angle are 128 m and 44 m respectively. Find the area of the plot in square metres.

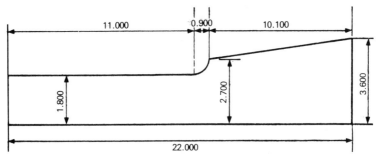

Figure 34

8 Calculate the total number of metres super in 12 hardwood planks, each 3.800 m long x 230 mm wide.

9 Find the number of metres run of 150 mm T & G boarding in 10 square metres of flooring.

10 The result of a survey of an area of building land is shown in Figure 35, all offsets being at right angles to the main line **AB**. Calculate the total area in square metres to the nearest m^2.

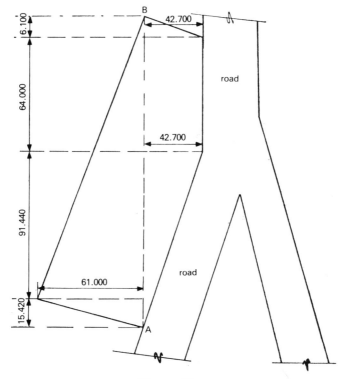

Figure 35

11   A certain type of flooring supplied and fixed in a rectangular area measuring 7.200 m x 4.200 m showed a total cost of £291.80. A different type of flooring supplied and fixed in a triangular area having a base of 3.600 m and a perpendicular height of 8.700 m cost a total of £166.80. Which is the least expensive flooring per square metre, and by how much?

12   What would be the width of a rectangular piece of insulating material required to wrap around a 3.050 m length of piping of 560 mm diameter, allowing 150 mm for a lap?

13   The floor of a circular tower has a diameter of 4.750 m. Find the floor area to the nearest square metre, and the length of the skirting to the nearest 10 mm, allowing 850 mm for a door opening.

14   Find the length of the hypotenuse of a right-angled triangle whose other two sides are 12 m and 16 m.

15   The end of a rope is attached to the top of a 16 m high mast and the other end fastened to a peg in the ground. If the rope is 20 m long, how far is the peg from the foot of the mast?

16   A box measuring 72 mm by 54 mm is to be divided into two compartments by a diagonal partition. What must be the length of the partition?

17   An embankment 30 m high spans 40 m horizontally. What is its actual sloping length?

18   A lean-to garage is 2.400 m wide and the roof is pitched so as to be 3.900 m high on one side and 2.100 m high on the other. What is the rafter-length of the roof?

19   A ladder 12.300 m long is to be placed so that it reaches a point 12 m high on a vertical wall. How far from the foot of the wall must its foot be stood?

20   A wall at the back of a theatre stage forms a semi-circle on plan with a diameter of 10.500 m. If its height is 4.500 m, find its surface area.

21   Figure 36 shows the end wall of a building having a semi-circular shell roof. The window has a semi-circular head. Calculate the total area of wall surface to be decorated if the reveals to the window are 150 mm wide. (The sill is not to be included.) Give your answer in m^2 to the nearest m^2 above.

22   The area of a square building plot is 36 m^2. What is its perimeter in metres?

23   A passage which is 1 m wide and 8 m long has its floor covered with 200 square tiles. What is the size of each tile in millimetres?

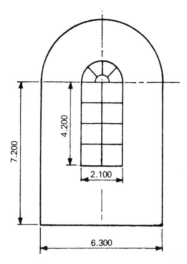

Figure 36

Properties of the circle

Now we shall consider some important properties of the circle. You will see marked in Figures 37 and 38 the following:

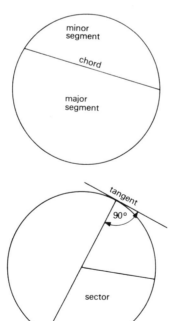

Figure 37

Figure 38

1   A *chord*, which is a straight line joining two points on the circumference but not passing through the centre,

2   A *minor segment*, which is the smaller portion of the circle cut off by a chord;

3   A *major segment*, which is the remainder of the circle;

4   A *tangent*, which is a straight line drawn outside the circle and touching its circumference at one point only. The radius at the point where a tangent touches the circumference is always at right angles to the tangent.

     Also shown in Figure 38 is a *sector*, which is a portion of the circle bounded by two radii and an arc. Radii is the plural of radius and an arc is a portion of the circumference. The sector shown is a *minor sector*, being less than half the circle. The remainder of the circle is a *major sector*.

Length of an arc

The sector shown in Figure 39 is a special one, the angle between the two radii being 90°, or a right angle. This means, of course, that the sector is exactly one quarter of the circle — called a *quadrant*. Clearly the length of the arc is one quarter of the circumference of the same circle.

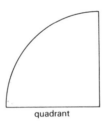

Figure 39

quadrant

     The number of degrees in a complete circle, one revolution, is 360. We can, therefore, think of the quadrant as $\frac{90}{360}$ of the circle, which is, of course, $\frac{1}{4}$ when cancelled down to its simplest terms. Thus, we can express the length of the arc of any sector in terms of the angle at the centre of the sector. Figure 40 gives some simple examples:

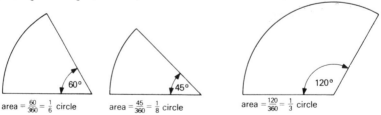

area $= \frac{60}{360} = \frac{1}{6}$ circle   area $= \frac{45}{360} = \frac{1}{8}$ circle   area $= \frac{120}{360} = \frac{1}{3}$ circle

Figure 40

If we write θ (theta) for the angle at the centre, we can say that, in general:

$$\text{Arc} = \frac{\theta}{360}\,\pi\,d$$

Example 9

What is the length of the arc of a sector which cuts off an angle of $42°$ at a radius of 50 mm?

$$\text{Arc} = \frac{\theta}{360}\,\pi d$$
$$= \frac{\overset{7}{4\cancel{2}}}{\underset{6}{3\cancel{60}}} \times 3.142 \times 50 \text{ mm}$$
$$= 18.32 \text{ mm}$$

No.	Log
7	0.8451
3.142	0.4972
5	0.6990
	2.0413
6	0.7782
18.32	1.2631

Note: Remember to cancel out before you evaluate by logs.

Area of sector

In the same way, the angle at the centre may be used to determine the sector's area. The area of each sector in Figure 40 must be the same fraction of the whole circle's area as the arc is of its circumference. So:

$$\text{Area of sector} = \frac{\theta}{360}\,\pi\,r^2$$

Example 10

A sheet of metal cut out to be formed into a cone has the shape of a circular sector of radius 1.500 m. The angle at the centre is $126°$. Find its area in square metres.

$$\text{Area of sector} = \frac{\theta}{360}\,\pi\,r^2$$
$$= \frac{126}{360} \times 3.142 \times 1.5 \times 1.5 \text{ m}^2$$
$$= 2.475 \text{ m}^2$$

No.	Log
1.5	0.1761
	2
1.5^2	0.3522
3.142	0.4972
126	2.1004
	2.9498
360	2.5563
2.475	0.3935

Area of segment

The best way to arrive at the area of a segment is to consider it as a sector from which a triangle has been removed (see Figure 41).

There is no *simple* formula for the area of a segment.

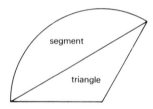

Figure 41

Example 11

The shape of a moulded strip is a segment of a circle of radius 100 mm. Its flat side is 50 mm and if the radii are drawn in they will make an angle at the centre of 29°.

Find the area of the strip's cross section. The area required is the shaded segment in Figure 42.

$$\text{Area of sector} = \frac{\theta}{360}\,\pi\,r^2$$
$$= \frac{29}{360} \times \pi \times 100 \text{ mm}^2$$
$$= 2531 \text{ mm}^2$$

No.	Log
29	1.4624
π	0.4972
	1.9596
360	2.5563
0.2531	1̄.4033

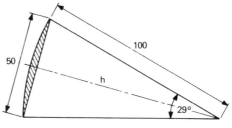

Figure 42

Note: 29 π ÷ 360 has been worked out by logs, and the result (0.2531) then multiplied by the 100² simply by moving the decimal point four places to the right.

Area of triangle

First, find the height, h, of the triangle by using the theorem of Pythagoras.

$$h^2 = 100^2 - 25^2$$
$$h = \sqrt{100^2 - 25^2}\ \text{mm}$$
$$= \sqrt{10000 - 625}\ \text{mm}$$
$$= \sqrt{9375}\ \text{mm}$$
$$= 96.83\ \text{mm (from tables)}$$

Area of triangle $= \frac{1}{2} \times 50 \times 96.83\ \text{mm}^2$
$= 25 \times 96.83\ \text{mm}^2$
$= 2420\ \text{mm}^2$

Area of segment $=$ sector – triangle
$= (2531 - 2420)\ \text{mm}^2$
$= 111\ \text{mm}^2$

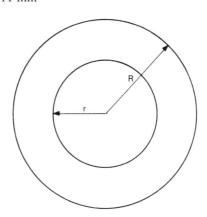

Figure 43

Area of an annulus

An annulus is a circular ring, the space contained between two concentric circles (circles on the same centre).

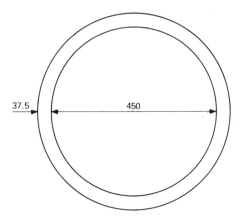

Figure 44

The area of the annulus is the difference between the areas of the two circles. Call the radius of the large circle R, and the radius of the small circle r.

Area of annulus $(A) = \pi R^2 - \pi r^2$

If we take out the common factor, π, this becomes:

$A = \pi (R^2 - r^2)$

This can be further simplified because $(R^2 - r^2)$ may be written as $(R - r)$ $(R + r)$, so the formula becomes:

$A = \pi(R - r)(R + r)$

It is a simple formula to use as shown in the following example.

Example 12

The wall thickness of a 450 mm diameter salt glazed stoneware drainage pipe is 37.5 mm. Find the cross sectional area of the stoneware.

This is, of course, an annulus, see Figure 44. The two radii are 225 mm and 262.5 mm.

Apply the formula:

$A = \pi (R - r)(R + r)$
$A = 3.142 (262.5 - 225)(262.5 + 225) \text{ mm}^2$
$\quad = 3.142 \times 37.5 \times 487.5 \text{ mm}^2$
$\quad = 57420 \text{ mm}^2$

No.	Log
3.142	0.4972
37.5	1.5740
487.5	2.6879
57420	4.7591

Sector of an annulus

This shape is often found in circular work and is shown in Figure 45.

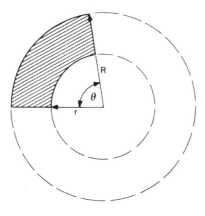

Figure 45

As in the case of the sector of a circle, the sector of an annulus is the same fraction of the whole annulus as the angle at the centre is of one complete revolution.

Area of sector of annulus $= \dfrac{\theta}{360}\, \pi\,(R - r)\,(R + r)$

Example 13

Two straight paths, 900 mm wide, meet at an angle of 120° and are joined by a curved portion of radius 3.600 m on the inside of the curve. What is the area of the curved portion of path?

Referring to Figure 46, C is the centre of the curved arc AB. The triangles ACX and BCX are right angled triangles and CX bisects the 120° angle AXB.

Thus, the angles AXC and BXC are each 60° and, therefore, the angles ACX and BCX are each 30°, making the angle at the centre of the sector 60°.

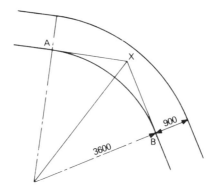

Figure 46

Area of curved path $= \dfrac{\theta}{360} \pi (R - r)(R + r)$

$= \dfrac{60}{360} \times 3.142 \times (4.5 - 3.6) \times (4.5 + 3.6) \ \text{m}^2$

$= \dfrac{1}{6} \times 3.142 \times 0.9 \times 8.1 \ \text{m}^2$

$= \dfrac{3.142 \times 7.29}{6} \ \text{m}^2$

$= 3.817 \ \text{m}^2$

No.	Log
3.142	0.4972
7.29	0.8627
	1.3599
6	0.7782
3.817	0.5817

Properties of the ellipse

Figure 47 shows an ellipse having a *major axis*, AB, and a *minor axis*, CD. The semi-major axis is marked a and the semi-minor axis b.

The area of the ellipse is given by the formula:

$A = \pi \, ab$

Its perimeter, or circumference, is given by:

$A = \pi (a + b)$

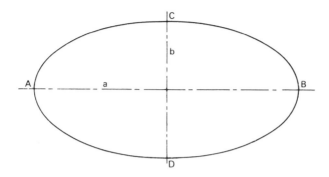

Figure 47

Example 14

Where it passes through a pitched roof a steel boiler flue pipe of 400 mm diameter makes an elliptical hole whose major axis is 500 mm. What is the area of roofing to be deducted? What is the length of jointing between the roof covering and the flue pipe?

First, the area:

$A = \pi ab$
  $= 3.142 \times 250 \times 200 \text{ mm}^2$
  $= 157100 \text{ mm}^2$

This may be required in square metres. Since there are 1000 mm in 1 m, there are $1000 \times 1000 \text{ mm}^2$ in 1 m^2.

The decimal point must therefore be moved six places to its left giving 0.1571 m^2.

The length of the joint between the pipe and the roof is the circumference of the ellipse.

$C = \pi (a + b)$
  $= 3.142(250 + 200) \text{ mm}$
  $= 3.142 \times 450 \text{ mm}$
  $= 1414 \text{ mm}$

Here are some more examples of problems involving the principles explained in this chapter.

Example 15

Figure 48 shows the shape of an end wall to a building whose roof is in the shape of a circular arc which cuts off an angle of 56° at its centre. Find the area of the shaded portion of wall.

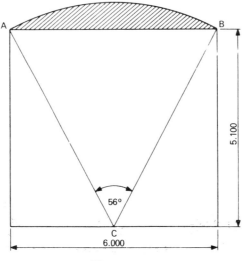

Figure 48

This is a segment of a circle. Its area will be the difference between the sector CAB and the triangle ABC.

$$\text{Sector CAB} = \frac{\theta}{360}\,\pi\,r^2$$

The radius CB can be found using the theorem of Pythagoras:

$$
\begin{aligned}
CB^2 &= 3^2 + 5.1^2 \\
&= 9 + 26.01 \\
&= 35.01 \\
CB &= \sqrt{35.01} \\
&= 5.917 \text{ m}
\end{aligned}
$$

$$\text{So, CAB} = \frac{56}{360} \times 3.142 \times 35.01 \text{ m}^2$$
$$= 17.11 \text{ m}^2 \text{ (by logs)}$$

No.	Log
56	1.7482
3.142	0.4972
35.01	1.5442
	3.7896
360	2.5563
17.11	1.2333

Triangle ABC $= \frac{1}{2} b \times h$
$= \frac{1}{2} \times 6 \times 5.1 \text{ m}^2$
$= 15.3 \text{ m}^2$

Area of segment   = sector – triangle
$= 17.11 - 15.3 \text{ m}^2$
$= 1.81 \text{ m}^2$

Example 16

A segmental arch has a radius of 1.500 m and its arc contains an angle of 106°. What is the length of the arc?

A segmental arch is one whose curve is a circular arc. The space between the springing line and the soffit (the curved underside of the arch) is a segment. The arch is shown in Figure 49.

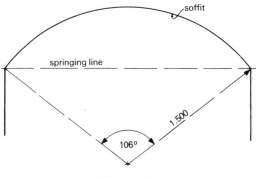

Figure 49

Arc $= \dfrac{\theta}{360} \pi d$

$= \dfrac{106}{360} \times \pi \times 3 \text{ m}$

$= 2.775 \text{ m}$

No.	Log
106	2.0253
π	0.4972
3	0.4771
	2.9996
360	2.5563
2.775	0.4433

Example 17

A semi-elliptical arch has a span of 1.800 m and a rise of 450 mm. What is the curved length of the arch soffit?

In the case of a semi-elliptical arch the span is the major axis of the ellipse and the rise is the semi-minor axis (see Figure 50).

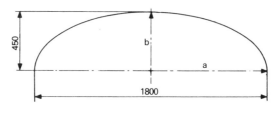

Figure 50

The length of the soffit is half the circumference of the ellipse, therefore:

Soffit length $= \frac{1}{2} \pi (a + b)$
$= \frac{1}{2} \times \pi \times (900 + 450)$ mm
$= \frac{1}{2} \times \pi \times 1350$ mm
$= \pi \times 675$ mm
$= 2120$ mm

No.	Log
π	0.4972
675	2.8293
2120	3.3265

Exercise 11

1   A circular water storage tank of diameter 9.000 m has a 750 mm wide walk round its edge. What is the area of the walk?

2   A curved boundary wall is to be built. On plan it is a circular arc of radius 9.000 m which cuts off an angle of $72°$ at its centre. What is its length?

3   The cross section of a handrail is shown in Figure 51. It is a major segment of a circle, radius 100 mm. Find its area.

4   A pipe flange has an internal diameter of 150 mm and an external diameter of 225 mm. It contains six 18 mm diameter bolt holes. What is the area of the face of the flange?

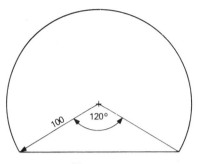

Figure 51

5   A concrete sun terrace is constructed as a semi-ellipse, with a length of 8.000 m and a breadth at the centre of 2.750 m. Find its area and the length of its curved edge.

6   A bend in a footpath 1.800 m wide consists of a curve of radius 2.700 m to the centre line of the path. The two straight lengths joined by the curve enclose an angle of 105°. Find the area of the curved portion of the path and the length of its curved edges.

7   A decorative brick fireplace has a segmental hearth, as shown in Figure 52. Find the area of the hearth and the length of its edge.

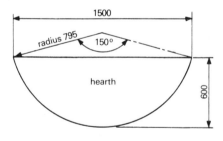

Figure 52

8   A circular bay window is constructed on the corner of a building, as shown in Figure 53. Its internal radius is 1.200 m and its width at its junction with the main walls is 1.550 m. The chord at this junction subtends (cuts off) an angle of 120° at the centre. Find the floor area in the bay and the length of curved skirting.

9   An elliptical manhole cover has a major axis of 700 mm and a minor axis of 500 mm. Find its perimeter and its area.

10 A paved rectangular area, 8.000 m by 6.000 m contains an elliptical pond. The minimum clearance between the pond and the edge of the paving is 900 mm on all four sides. Find the area of paving in square metres.

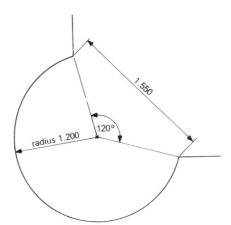

Figure 53

11   A decorative floor pattern consists of an ellipse containing a central annulus and two circular sectors each with an angle of 120°, as shown in Figure 54. The pattern is in two colours as indicated by the shading. The ellipse is 4.800 m by 2.700 m, the inner diameter of the annulus and the diameter of the sectors is 1.200 m and the outer diameter of the annulus is 2.100 m. Find the area of each colour.

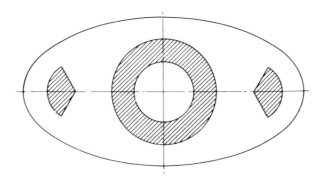

Figure 54

12   Find the areas and perimeters of the following:
(a)   An annulus whose radii are 85.3 mm and 14.7 mm.
(b)   A sector of radius 1.575 m containing an angle of 72°.
(c)   A segment on a chord of length 3.000 m which subtends an angle of 90° at the centre.
(d)   An ellipse, major axis 61.4 mm, minor axis 38.6 mm.

7 Triangles and regular polygons

In Chapter 6 we considered certain properties of triangles, particularly the right-angled triangle and the relationship between its sides known as the theorem of Pythagoras. You will have noticed that this theorem has been used many times in examples. We shall now deal with triangles in more detail.

The sum of the interior angles of a triangle is 180°, or two right angles.

If one side of a triangle is produced the angle made with the adjacent side (∠ ACD in Figure 55) is called an *exterior* angle. Since this angle, together with the interior angle (∠ ACB) makes 180°, it follows that it is equal to the sum of the other two interior angles (since they also, together with ∠ ACB, make 180°). That is, an exterior angle is equal to the sum of the two interior opposite angles.

In Figure 55, ∠ ACD = ∠ ABC + ∠ BAC.

In any triangle the side opposite the greatest angle is the longest side and the side opposite the smallest angle is the shortest side.

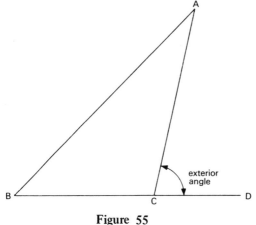

Figure 55

In a right-angled triangle the side opposite the right angle is the longest side, called the hypotenuse. The other two angles have a sum of 90°, or one right angle, and are called *complementary* angles.

Two angles whose sum is 180° are called *supplementary* angles.

When all the angles in a triangle are acute (less that 90°) it is called an *acute-angled triangle*.

When one angle is obtuse (greater than 90°) it is called an *obtuse-angled triangle*.

When all its sides are unequal in length a triangle is called a *scalene triangle*.

If it has two equal sides it is called an *isosceles triangle* and, because it has two equal sides, it has two equal angles.

When all three sides and, therefore, all three angles of a triangle are equal it is an *equilateral triangle*.

These triangles are all shown in Figure 56.

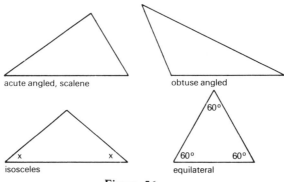

Figure 56

Two triangles with which you should be particularly familiar are the 45° right-angled triangle and the 30°/60° right-angled triangle. These are the shapes of your set squares.

Figure 57 shows a 45° right-angled triangle. The angles B and C are each 45°. The sides b and c are therefore equal, and, as well as being a right-angled triangle, it is an isosceles triangle. If we make sides b and c each one unit long and use the theorem of Pythagoras, we have:

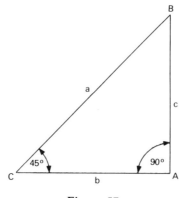

Figure 57

$$a^2 = b^2 + c^2$$
$$= 1^2 + 1^2$$
$$= 1 + 1$$
$$a = \sqrt{2}$$

This is an important result and tells us that the sides of a 45° right-angled triangle are in the constant relationship $1 : 1 : \sqrt{2}$. Reference to the square root tables will show you that $\sqrt{2} = 1.414$. Try to memorize this.

If the equal sides of such a triangle are 10 units, then the long side is $10\sqrt{2} = 10 \times 1.414 = 14.14$ units.

Example 1

What is the length of the hypotenuse of a 45° right-angled triangle whose equal sides are 15 cm?

Hypotenuse $= 15\sqrt{2}$ cm
$= 15 \times 1.414$ cm
$= 21.21$ cm

Example 2

The hypotenuse of a 45° right-angled triangle is 600 mm long. What are the lengths of the other two sides?

They are, of course, equal and their length is:

$$\frac{600}{\sqrt{2}} \text{ mm}$$

There is a method of simplifying this fraction to make it easy to evaluate. Multiply the numerator and the denominator by $\sqrt{2}$, then we have:

$$\frac{600 \times \sqrt{2}}{\sqrt{2} \times \sqrt{2}}$$

The denominator now becomes 2 because:

$$\sqrt{2} \times \sqrt{2} = 2$$

The fraction becomes:

$$\frac{600\sqrt{2}}{2}$$
$$= 300\sqrt{2}$$
$$= 300 \times 1.414$$
$$= 424.2 \text{ mm}$$

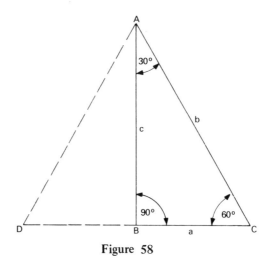

Figure 58

Figure 58 shows a 30°/60° right-angled triangle ABC. When the dotted triangle ABD is drawn in it shows that ABC is in fact one half of an equilateral triangle.

Now, if we call the side a one unit, the side b will be two units long. Using the theorem of Pythagoras:

$$c^2 = b^2 - a^2$$
$$= 2^2 - 1^2$$
$$= 4 - 1$$
$$= 3$$
$$c = \sqrt{3}$$

So we see that in a 30°/60° right-angled triangle the sides are in the constant relationship $1 : 2 : \sqrt{3}$. Note that the shortest side is represented by 1, the longest side (hypotenuse) by 2 and the intermediate side by $\sqrt{3}$.

$\sqrt{3} = 1.732$, another value well worth memorizing.

If the short side of a 30°/60° right-angled triangle is 10 units then the hypotenuse will be 20 units and the intermediate side will be:

$$10\sqrt{3} = 10 \times 1.732$$
$$= 17.32 \text{ units}$$

Example 3

In a 30°/60° right-angled triangle the short side is 5 cm. Find the other two sides.

The longest side is twice the shortest side, that is:

Hypotenuse = 2 x 5

= 10 cm

The intermediate side is $\sqrt{3}$ times the short side, that is:

Third side = $\sqrt{3}$ x 5

= 1.732 x 5

= 8.66 cm

Isosceles triangles occur frequently in constructional work. The following examples illustrate some applications.

Example 4

The legs of a scaffold trestle are each 1.800 m long and open to enclose an angle of 40° when the feet are 1.230 m apart. What angles do the legs make with the floor, and what is the height of the trestle? (See Figure 59.)

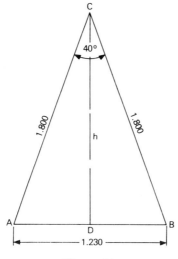

Figure 59

Since the legs are equal, the triangle is isosceles and, therefore, the base angles are equal. As the sum of all three angles is 180°:

∠ A + ∠ B = 180° – 40°

= 140°

and ∠ A = ∠ B

= 70°

In the triangle ACD:

$$AD = \frac{1.230}{2} m$$
$$= 0.615 \text{ m}$$
$$h^2 = 1.8^2 - 0.615^2$$
$$= 3.240 - 0.3782$$
$$= 2.862$$
$$h = \sqrt{2.862} \text{ m}$$
$$= 1.692 \text{ m}$$

Example 5

A span roof is pitched at 35°. It has a span of 5.400 m and a rise of 1.900 m. Find the length of the rafter and the angles at which it must be cut at the ridge and the eaves.

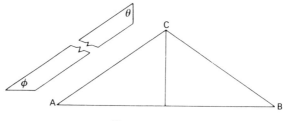

Figure 60

Figure 60 shows the problem diagrammatically. The triangle ABC is isosceles. The required angles have been labelled θ (theta) and ϕ (phi). These are both letters of the Greek alphabet commonly used to represent angles. The length of the rafter is AC (or BC).

By inspection the angle at the eaves, ϕ, is the pitch of the roof, that is, 35°.
The angle at the ridge, θ, is:

$\frac{1}{2} [180° - (2 \times 35°)]$
$\frac{1}{2} (180° - 70°)$
$\frac{1}{2} (110°) = 55°$

or, more simply, $(90° - 35°) = 55°$
For the length of the rafter we use the theorem of Pythagoras again.

$$AC^2 = 2.7^2 + 1.9^2$$
$$= 7.29 + 3.61$$
$$= 10.90$$
$$AC = \sqrt{10.9} \text{ m}$$
$$= 3.302 \text{ m}$$

Example 6

Figure 61 shows the framing of a roof truss. This type of truss is called a *scissors* truss. EF is horizontal. Find the angles AED and ADE, and the lengths of AC and AD. The span is 9 m.

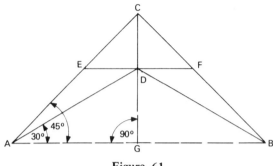

Figure 61

Join AB and produce CD downwards to meet AB at G. The angle thus made, $\angle$ AGC is $90°$. Therefore:

$\angle$ ADG $= 60°$
$\angle$ ADE $= 90° - 60° = 30°$
$\angle$ EAD $= 45° - 30° = 15°$

In the triangle AED:

$\angle$ AED $= 180° - (\angle$ EAD $+ \angle$ ADE)
$= 180° - (15° + 30°)$
$= 180° - 45°$
$= 135°$

Triangle ACG is a $45°$ right-angled triangle and its two short sides are each 4.500 m. Therefore:

AC $= 4.5 \sqrt{2\,m}$
$= 4.5 \times 1.414$ m
$= 6.363$ m

Triangle ADG is a $30°/60°$ right-angled triangle. The sides are therefore in the relationship $1 : 2 : \sqrt{3}$, and the side AG, 4.5 m long, corresponds to $\sqrt{3}$.

The short side, DG $= \dfrac{4.5}{\sqrt{3}}$ m

$= \dfrac{4.5 \times \sqrt{3}}{\sqrt{3} \times \sqrt{3}}$ m (see Example 2)

$= \dfrac{4.5 \sqrt{3}}{3}$ m

$$= 1.5 \sqrt{3} \text{ m}$$
$$= 2.598 \text{ m}$$

The long side, AD $= 2 \times 2.598$ m
$$= 5.196 \text{ m}$$

Similar triangles

Triangles having the same *shape* (that is, equal angles) but not, necessarily, the same *size*, are called *similar triangles*. For example, all $30°/60°$ right-angled triangles are similar, as are all $45°$ right-angled triangles.

As we have already seen in the above-mentioned two cases, the sides of similar triangles are proportional.

Example 7

Figure 62 shows a triangular frame to carry a partition under a staircase. Calculate the length of the vertical members DE and FG.

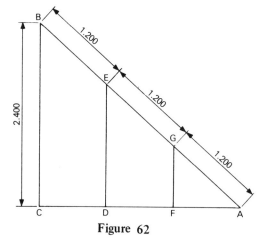

Figure 62

In the diagram the triangles ABC, AED and AGF are similar. Since the sides of similar triangles are proportional, it follows that:

$$\frac{BC}{AB} = \frac{ED}{AE} = \frac{GF}{AG}$$

First, to find ED.

$$\frac{ED}{AE} = \frac{BC}{AB}$$

$$\frac{ED}{2.4} = \frac{2.4}{3.6}$$

$$ED = \frac{2.4 \times 2.4}{3.6}$$

$$= \frac{5.76}{3.6}$$

$$= 1.600 \text{ m}$$

Now, to find GF.

$$\frac{GF}{AG} = \frac{BC}{AB}$$

$$\frac{GF}{1.2} = \frac{2.4}{3.6}$$

$$GF = \frac{1.2 \times 2.4}{3.6}$$

$$= \frac{2.88}{3.6}$$

$$= 0.800 \text{ m}$$

Congruent triangles

When similar triangles are also of the same size, that is, equal in all respects, they called *congruent triangles*.

Regular polygons

A polygon is a figure having five or more straight sides. (A four-sided figure is a quadrilateral.) If the sides of a polygon are equal then the angles will also be equal and the polygon is said to be *regular*. A regular polygon may be divided up into a number of congruent, isosceles triangles; some regular polygons are shown in the following figures.

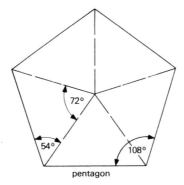

pentagon

Figure 63

Figure 63 is a *pentagon*, having five sides. The five angles at the centre are each,

$$\frac{360}{5} = 72°$$

The base angles of the five isosceles triangles are therefore:

$$\frac{180 - 72}{2} = \frac{108}{2}$$
$$= 54°$$

and the five angles between the sides of the pentagon are each 108°.

Figure 64 is a *hexagon*, having six sides. This contains six equilateral triangles, all the angles being 60°. The angles between the sides of the hexagon are, therefore, each 120°.

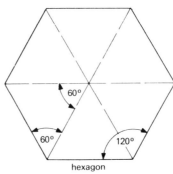

hexagon

Figure 64

Figure 65 shows a *heptagon*, which has seven sides. The seven angles at the centre are each,

$$\frac{360°}{7} = 51\tfrac{3}{7}°$$

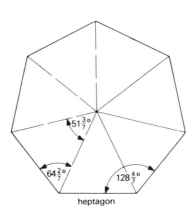

heptagon

Figure 65

The base angles of each of the seven triangles are each:

$$\frac{180° - 51\frac{3}{7}°}{2} = \frac{128\frac{4}{7}°}{2}$$

$$= 64\frac{2}{7}°$$

The angles between the sides of the heptagon are, therefore, $128\frac{4}{7}°$. Other regular polygons are:

Octagon   – eight sides
Nonagon   – nine sides
Decagon   – ten sides
Undecagon   – eleven sides
Duodecagon   – twelve sides

Example 8

The plan of an octagonal room is shown in Figure 66. Calculate the area of the floor and the angle at each corner of the room.

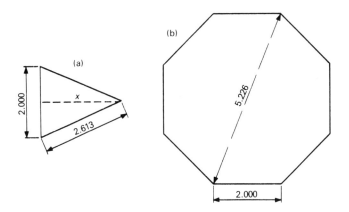

Figure 66

Considering one of the eight triangles of which the octagon is formed (see Figure 66 (a)).

$$x^2 = 2.613^2 - 1^2$$
$$= 6.828 - 1$$
$$= 5.828$$
$$x = \sqrt{5.828}$$
$$= 2.414 \text{ m}$$

Area of octagon = 8 x area of each triangle
$$= 8 \times \tfrac{1}{2} (2 \times 2.414) \, m^2$$
$$= 19.312 \, m^2$$
Each of the eight angles at the centre
$$= \frac{360}{8}$$
$$= 45°$$

The other two angles of each triangle are therefore:
$$\frac{180 - 45}{2} = \frac{135}{2}$$
$$= 67\tfrac{1}{2}°$$

Corner angle $= 2 \times 67\tfrac{1}{2}°$
$$= 135°$$

Exercise 12

1   What are the complementary angles to the following:
   (a)   47°
   (b)   28°
   (c)   15°
   (d)   24°
   (e)   8°

2   Write down the supplementary angles to the following:
   (a)   140°
   (b)   96°
   (c)   53°
   (d)   18°
   (e)   124°

3   In a triangle ABC, angle C is 40° and the exterior angle at B, formed by producing CB, is 120°. What are the other two interior angles of the triangle?

4   Draw the following triangles:
   (a)   a = 7.5 cm, B = 30°, C = 45°. What is the third angle? Measure the sides b and c.
   (b)   a = 6 cm, b = 8 cm, c = 10 cm. Without measurement state the value of the largest angle.
   (c)   a = b = 7 cm, c = 75°. What are the angles A and B?

5   The long side of a 45° set square is 25 cm. Find the lengths of its other two sides.

6 A prop, against a vertical partition, makes an angle of 60° with the floor. The prop is 2.400 m long. What is the height of its top above floor level and the distance of its foot from the partition?

7 Figure 67 shows a framed roof truss. AG, GF, FE, GC and CF are equal in length AB, BC, CD, and DE are also equal. Find the length of all the members and the angles between them.

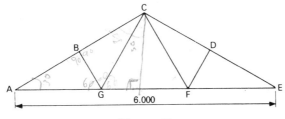

Figure 67

8 A north-light roof truss is shown in Figure 68. Find its rise (height from eaves to ridge) and the lengths of its rafters.

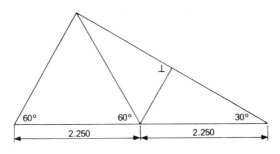

Figure 68

9 It is required to brace a 1.200 m square frame by inserting a diagonal member. What will be the length of the brace?

10 The diagonal of a rectangular sheet of plywood is twice the length of its short side, which is 800 mm. What is the area of the sheet?

11 The equal legs of a step ladder are 1.350 m long. They open to a maximum width at the feet of 1.350 m. What is the height of the top?

12 Two guy ropes on opposite sides of a flag pole 17.320 m high, form, with the ground, an isosceles triangle. The ropes are fixed at the top of the pole and at points 10.000 m from its base. Find the angle that the ropes make with the ground and the length of each rope.

13 A span roof has a pitch of $42\frac{1}{2}°$. Its rise over a span of 7.200 m is 3.300 m. Find the rafter length and the angle at which it must be cut at the ridge.

8 Volumes and surface areas of solids

Volume is defined as 'space taken up' and must not be confused with mass which is 'the amount of matter contained'. If you think of two equal sized cubes of lead and aluminium you will realize that, although they have equal *volumes*, the lead has a much larger *mass* than the aluminium. Your science lecturer will explain this to you in greater detail.

The *volume* of an object is found by multiplying three linear dimensions together with the result appearing as a *cubic* measurement, that is, cubic metres, cubic centimetres, cubic millimetres, etc.

Just as we saw in Chapter 6 that the *square* was the unit of area, we must now recognize the *cube* as the unit of volume.

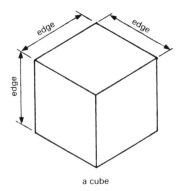

a cube

Figure 69

The volume of a cube = edge x edge x edge. If we denote edge by the letter e and volume by the letter V we can write:

$V = e^3$ (*V* equals *e* cubed)

It is important at this stage to have a clear picture of the connection between the *linear* (length), *superficial* (area) and *cubic* (volume) units. This is best clarified pictorially as shown in Figure 70.

Note: The units are squared or cubed as well as the quantity.

Units of linear, superficial and cubic measure are quite different and there is no question of conversion from one to another. You can refer to the number of square millimetres in a square metre but do not even consider the idea of expressing so many cubic metres in square metres.

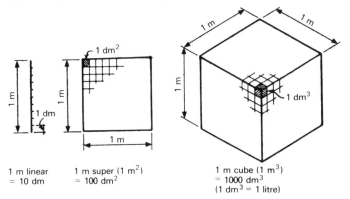

Figure 70

The *weights* of building materials are expressed as so many kilograms per cubic metre (kg/m^3). For example:

Structural concrete   weighs 2400 kg/m^3
Steel   weighs 7830 kg/m^3
White pine   weighs   530 kg/m^3

In each case the figure given is the mass of one unit of volume of the material. This is referred to as *density*. (That is, density is strictly *mass* per unit volume.)

If we have a volume of material we can easily find its mass by multiplying volume by density.

Example 1

What is the mass of a 150 mm concrete cube if the density of the concrete is 2400 kg/m^3?
Volume =   150 x 150 x 150 mm^3
       or   0.15 x 0.15 x 0.15 m^3
(each edge dimension has been divided by 1000 to convert from mm to m)
   This gives a volume of 0.003375 m^3
Weight = volume x density
       = 0.003375 x 2400 kg
       = 8.1 kg

Prisms

A prism is a solid which has the same cross-section throughout its length and its top parallel to its base.

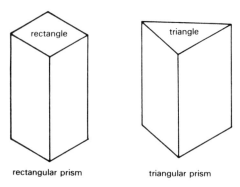

rectangular prism   triangular prism

Figure 71

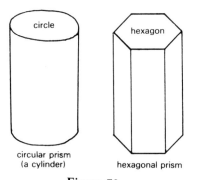

circular prism
(a cylinder)   hexagonal prism

Figure 72

Figures 71 and 72 are examples of *right* prisms because the side faces are rectangular (this is even true for the cylinder if you develop the surface).

The *volume* of a right prism is the area of one end multiplied by the height of the prism or, *volume of prism = area of cross-section × height*.

To find the total *surface area* of a prism, *add* the areas of all its faces.

Where the prism is *inclined*, the formula for the volume holds good provided you remember that the *vertical height* of the prism must be taken.

Example 2

A concrete drive to a garage is shown as a plan in Figure 73. How many cubic metres of concrete are required to construct the drive?

This drive is in the shape of an inclined rectangular prism of the following dimensions:

Volume of concrete = area of cross-section × vertical height
$$= (3.000 \times 0.300) \times 10.500 \text{ m}^3$$
$$= 9.45 \text{ m}^3$$

Example 3

How many bricks would you require to build a wall 225 mm thick, measuring 8.400 m long and 1.800 m high. (*Note:* Take the dimensions of a brick with mortar to be 225 mm x 112.5 mm x 75 mm.)

First find the volume of the wall in cubic metres:

Volume of wall = 8.4 x 1.8 x 0.225 m^3
            = 3.40 m^3

Now the volume of one brick with mortar (also in m^3):

Volume = 0.225 x 0.1125 x 0.075 m^3
       = 0.0019 m^3 (to four significant figures)

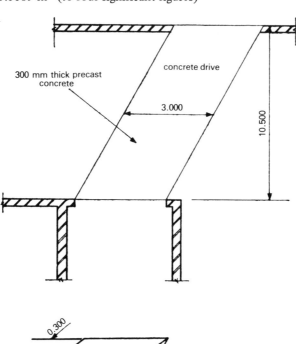

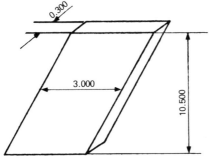

Figure 73

The number of bricks required will be the number of times the volume of one brick can be divided into the total volume.

Number of bricks = 3.40 ÷ 0.0019
= 1790 bricks

Note: This answer gives the actual number of bricks contained in the wall and in practice you would probably add a percentage of this total in order to make an allowance for wastage in cutting, etc.

Example 4

How many cubic metres of timber are contained in a floor consisting of 100 joists each 2.850 m long x 75 mm x 225 mm in cross-section. (Give your answer to three significant figures.)

Total volume of one joist is the area of cross-section multiplied by the length.

Volume of one joist   = 2.850 x 0.075 x 0.225 m^3
= 0.048 m^3
Volume of 100 joists   = 0.048 x 100 m^3
= 4.8 m^3

Example 5

If the density of a certain building stone is 2080 kg/m^3, what is the volume of 1 tonne of the stone? (Give your answer correct to the first place of decimals.)
Note: 1 tonne = 1000 kg. This is the 'metric tonne'.

The answer will obviously be equal to the number of times 2080 kg may be divided into 1 tonne.

Volume $= \frac{1000}{2080} m^3$
= 0.48 m^3 (this is, approximately $\frac{1}{2}$ m^3)

Example 6

An artificial stone plinth course is cast in 900 mm lengths with the cross-section shown in Figure 74. Find the volume of each block, giving your answer in cubic metres.

The block has a cross-section which is best regarded as a rectangle minus a triangle. The net area of this times the length will give its volume.

Cross-sectional area = rectangle ABCD – triangle DEF
= (300 x 150) – $\frac{1}{2}$ (100 x 75) mm^2

$$= 45000 - 3750 \text{ mm}^2$$
$$= 41250 \text{ mm}^2$$
Volume $\qquad = 41250 \times 900 \text{ mm}^3$
$$= 37125000 \text{ mm}^3$$

Since there are $1000 \times 1000 \times 1000 \text{ mm}^3$ to 1 m^3 the decimal point must be moved 9 places to the left, giving 0.037 m^3.

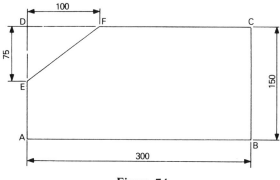

Figure 74

Example 7

A bar of steel has a hexagonal cross-section as shown in Figure 75 and a length of 4.500 m. If steel weighs 7800 kilograms per cubic metre, what is the weight of the bar?

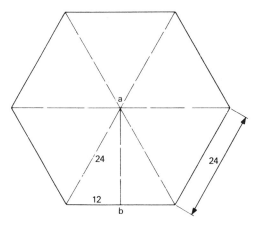

Figure 75

The bar is a hexagonal prism. A hexagon consists of six equilateral triangles and its area is, therefore, six times the area of one of them. The perpendicular height ab of the triangle thus marked divides it into two right-angled triangles and, using the theorem of Pythagoras, we have:

$ab^2 = 24^2 - 12^2$
$ab = \sqrt{576 - 144}$
$= \sqrt{432}$
$= 20.78$ mm

Area of each triangle $= \frac{1}{2} bh$
$= \frac{1}{2} \times 24 \times 20.78$ mm^2
$= 249.36$ mm^2

Area of hexagon $= 6 \times 249.36$ mm^2
$= 1496.16$ mm^2 (say 1496)

Volume of bar $= \dfrac{1496 \times 4.5}{10^6}$ m^3

Weight of bar $= \dfrac{1496 \times 4.5 \times 7800}{10^6}$ kg

$= 52.5$ kg

No.	Log
1496	3.1749
4.5	0.6532
7800	3.8921
	7.7202
10^6	6.0000
52.50	1.7202

Note: Dividing by 10^6 changes mm^2 to m^2 since there are (1000 x 1000) mm^2 to 1 m^2.

Example 8

An embankment is 7.600 m wide at its base and 1.800 m wide at its top, with equal sloping sides of length 3.500 m. How many cubic metres of earth are contained in a kilometre of its length? Its cross-section is shown in Figure 76.

You will recall that this shape is a trapezium and that the area of a trapezium is obtained by, half the sum of the parallel sides times the perpendicular distance between them.

In triangle ABC:

$AB = \frac{1}{2} (7.6 - 1.8)$
$= 2.9$ m

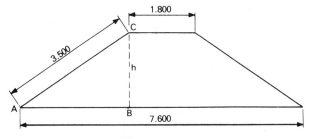

Figure 76

If we use the theorem of Pythagoras:

$h^2 = 3.5^2 - 2.9^2$
$h = \sqrt{3.5^2 - 2.9^2}$ m
$= \sqrt{12.25 - 8.41}$ m
$= \sqrt{3.84}$ m
$= 1.96$ m

Area of trapezium $= \dfrac{7.6 + 1.8}{2} \times 1.96$ m^2

$= 4.7 \times 1.96$ m^2
$= 9.212$ m^2

Volume of embankment $= 9.212 \times 1000$ m^3
(1 km = 1000 m)

$= 9212$ m^3

No.	Log
4.7	0.6721
1.96	0.2923
9.212	0.9644

Exercise 13

1   What volume of concrete is contained in an 8.400 m length of foundation, the dimensions of which are shown in Figure 77?

2   What total weight of lead weighing 34.230 kg per m^2 would be required to cover the cheeks and top of the dormer window shown in Figure 78, allowing 10% for laps?

3   What is the actual capacity in litres of a cold water storage tank measuring 675 mm square on plan, and 750 mm deep?

4   A 1.500 m wide drain trench is to be excavated in level ground for a length

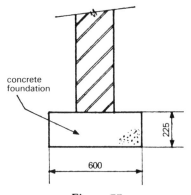

Figure 77

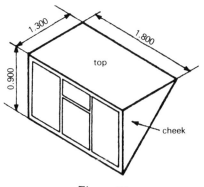

Figure 78

of 240 m and to a fall of 1 in 60. If the starting depth is 1.000 m, calculate in m³ the total volume of earth to be excavated making no allowance for bulking.

5   Find the cost at £150 per m³ of 65 m run of 75 mm x 50 mm wall plate.

6   If mass concrete weighs 2240 kg/m³, what is the weight of a rectangular concrete pillar measuring 450 mm square on plan and 1.950 m high?

7   If a bricklayer's output is approximately 600 bricks laid per day, how long would you allow for two bricklayers to build a 225 mm boundary wall, 100 m long and 1.800 m high? (Allow 4 courses to every 300 mm.)

8   Figure 79 shows the cross-section of a trench excavation which has a length of 18.000 m. When excavated the earth increases in bulk by 25%. How many cubic metres are to be carted away.

9   A concrete block used to support piping is shown in Figure 80. What is the volume of the block in cubic metres and what is its total surface area?

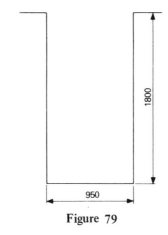

Figure 79

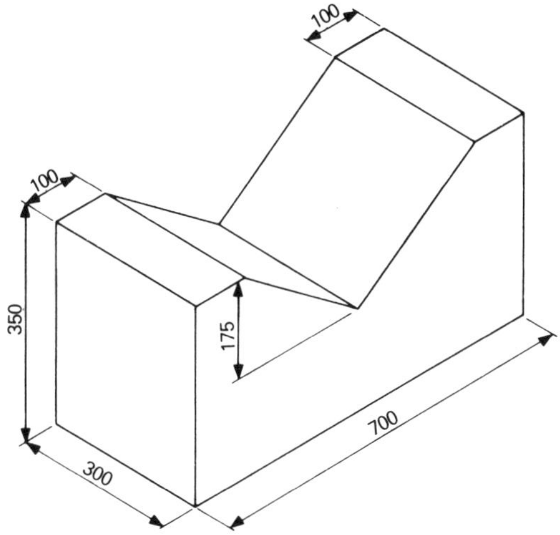

Figure 80

10   A building is rectangular on plan and measures 14.400 m long and 6.600 m wide. Its roof, which has hipped ends, is pitched at 45°. What volume is enclosed by the roof? (*Note*: The two hipped ends together would form a square pyramid, the portion between being a triangular prism.)

11   A reinforced concrete pile is rectangular in cross-section, measuring 350 mm by 300 mm. Its total length is 4.800 m and one end is tapered off to a point over a length of 600 mm. The material weighs 2400 kilograms per cubic metre. Find the weight of the pile.

Cylinder

The volume, V, of a cylinder, being the area of its cross-section (a circle) times its height, is given by the formula:

$V = \pi r^2 h$

h being the height and r the radius.

The surface area of a cylinder may also be expressed in a formula. If the curved surface is developed (opened out flat) it becomes a rectangle whose length is equal to the cylinders circumference and whose breadth is the height of the cylinder (see Figure 81).

The area of the rectangle is therefore $2\pi r \times h$. That is:

curved surface of cylinder = $2\pi rh$

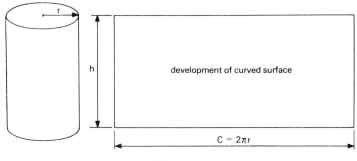

Figure 81

To obtain the total surface area of the cylinder it is necessary to add to this curved surface the areas of the two end circles. Writing S for this total surface we have:

$S = 2\pi rh + 2\pi r^2$

You will see that $2\pi r$ occurs in both terms of this expression, and is therefore a 'common factor'. The formula can be simplified as:

$S = 2\pi r(h + r)$

Example 9

Find the capacity in litres and total surface in square metres of a copper cylinder having a diameter of 525 mm and a height of 825 mm. (Capacity simply means volume contained.)

Capacity of cylinder:

$V = \pi r^2 h$
$\quad = \pi \times (262.5)^2 \times 825 \text{ mm}^3$
$V = \dfrac{\pi \times (262.5)^2 \times 825}{10^6} \text{ litre}$
$\quad = 178.6 \text{ litre}$

No.	Log
262.5	2.4191
$(262.5)^2$	4.8382
825	2.9165
π	0.4972
	8.2519
10^6	6.0000
178.6	2.2519

Note: 1 litre is a cubic decimetre and, since there are 100 mm to a decimetre, there are 100^3 mm^3 to 1 litre $(100^3 = 10^6)$.

Total surface:

$S = 2\pi r(h + r)$
$= 2 \times \pi \times 262.5 \times (825 + 262.5)$ mm^2
$S = \dfrac{2 \times \pi \times 262.5 \times 1087.5}{10^6}$ m^2
$= 1.795$ m^2

No.	Log
2	0.3010
π	0.4972
262.5	2.4191
1088	3.0366
	6.2539
10^6	6.0000
1.795	0.2539

Note: Since there are 1000 mm to 1 m, there are 1000^2 mm^2 to 1 m^2 $(1000^2 = 10^6)$.

Hollow cylinder

This is, as you see in Figure 82, a tube, its cross-section being an annulus.
    Since the area of an annulus is given by:

$A = \pi(R - r)(R + r)$

the volume of the hollow cylinder will be obtained if we multiply this area by the height, h.
    The volume is therefore:

$V = \pi h(R - r)(R + r)$

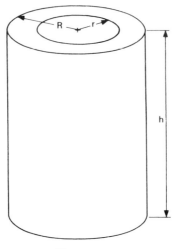

Figure 82

Example 10

Find the volume of material in the walls of a pipe 1.200 m long and having internal and external diameters of 60 mm and 90 mm. Give the answer in mm^3.
    Note that the diameters are given. We need the radii.

R = 45 mm
r = 30 mm

$V = \pi h\,(R - r)\,(R + r)$
    $= \pi \times 1200 \times (45 - 30) \times (45 + 30)$ mm^3
    $= \pi \times 1200 \times 15 \times 75$ mm^3
    $= 4242000$ mm^3
    $= 4.242 \times 10^6$ mm^3

No.	Log
π	0.4972
1200	3.0792
15	1.1761
75	1.8751
4242000	6.6276

Note: The large number of mm^3 has been divided by a million and the million then expressed in index form as 10^6. This is a convenient way of expressing large numbers and is known as standard form.

The sphere

The volume of a sphere is found, by using the formula:

$V = \frac{4}{3}\pi r^3$

Its surface area is given by:

$S = 4.\pi r^2$

Example 11

12 mm diameter steel balls are to be used as balance weights. How many of them will be needed to balance a steel cylinder 25 mm in diameter and 75 mm high?

Since they are made of the same material, a comparison of their volumes will give the same ratio as a comparison of their weights. It is not, therefore, necessary to know the weight of the steel in order to solve the problem.

Volume of cylinder:

$$V_c = \pi r^2 h$$
$$= \pi \times (12.5)^2 \times 75 \text{ mm}^3$$
$$= 11710 \,\pi \text{ mm}^3$$

Volume of one ball:

$$V_s = \frac{4}{3}\pi r^3$$
$$= \frac{4}{3} \times \pi \times 6^3 \text{ mm}^3$$
$$= 288 \,\pi \text{ mm}^3$$

No.	Log
12.5	1.0969
(12.5)	2.1938
75	1.8751
11710	4.0689

Now, the number of balls required will be the result of dividing the volume of the cylinder by the volume of the sphere.

Number required $= \dfrac{11710\,\pi}{288\,\pi}$

$= 40.69$ (say 41 balls)

No.	Log
11710	4.0689
288	2.4594
40.69	1.6095

Pyramids

Like prisms, pyramids take their names from the shape of their bases, so we may have square, rectangular or hexagonal pyramids, to quote only three. These are shown in Figure 83.

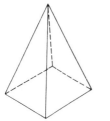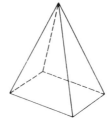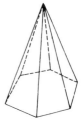

Figure 83

The volume of a pyramid is always *one third the volume of the enclosing prism* (this means a prism on the same base and with the same height).

The surface area of a pyramid will be obtained by adding together the base area and the areas of the triangles forming the sloping sides.

Example 12

A roof on a 6.000 m square building is pitched at 45°, forming a square pyramid. Find the volume of space enclosed by the roof, and the area of its surface.

As the pitch of the roof is 45°, the height of the roof will be equal to half the span that is, 3.000 m.

First find the volume:

$V = \frac{1}{3}a^2 h$ (where a is the side)

$ = \frac{1}{3} \times 6 \times 6 \times 3 \text{ m}^3$

$ = 36 \text{ m}^3$

The roof area is the sum of four equal triangles. The base of each triangle is 6.000 m and the perpendicular height is the slant height, S in Figure 84.

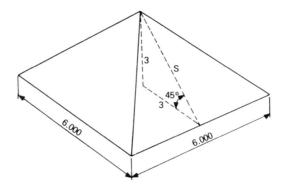

Figure 84

$$S^2 = 3^2 + 3^2$$
$$S = \sqrt{9+9} \text{ m}$$
$$= \sqrt{18} \text{ m}$$
$$= 4.243 \text{ m}$$

Area of roof surface $= 4 \times \frac{6}{2} \times 4.243 \text{ m}^2$
$= 50.916 \text{ m}^3$

The cone

A special pyramid, which has a circular base.

The same rule applies for its volume, that is, one-third the base area times the height. Therefore:

$$V = \tfrac{1}{3} \pi r^2 h$$

The curved surface of a cone is given by the formula:

$$S = \pi r l$$

where l is the slant height as shown in Figure 85. This may be found by using the theorem of Pythagoras if the radius and the height are known.

For the total surface of a cone, the area of the base must be added to the curved surface area, which gives:

$$S = \pi r l + \pi r^2$$

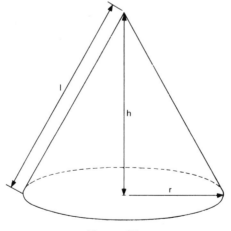

Figure 85

Again we have a common factor, which is πr, and if we factorize, the formula becomes:

$$S = \pi r (l + r)$$

Example 13

Eight tonnes of excavated earth are tipped into a conical heap of base diameter 3.600 m and maximum height 1.200 m. How many cubic metres of earth does the heap contain, and what is its bulk density in kilograms per cubic metre?

First, find the volume:

$V = \frac{1}{3} \pi r^2 h$
$= \frac{1}{3} \pi \times 1.8^2 \times 1.2 \ \text{m}^3$
$= 4.072 \ \text{m}^3$

No.	Log
1.8	0.2553
$(1.8)^2$	0.5106
1.2	0.0792
π	0.4972
	1.0870
3	0.4771
4.072	0.6099

Now, find the density:

$D = \dfrac{\text{weight}}{\text{volume}}$

$= \dfrac{8 \times 1000}{4.072} \ \text{kg/m}^3$

$= 1964 \ \text{kg/m}^3$

No.	Log
8000	3.9031
4.072	0.6099
1964	3.2932

Exercise 14

1   A cylindrical sash weight measuring 40 mm in diameter and 350 mm long weighs 3.2 kg.
   (a)   What is its density (weight in kg per m³)?
   (b)   Calculate the weight of a larger sash weight of similar material measuring 50 mm in diameter and 450 mm long.

2   Calculate the number of cubic metres of brickwork in a 3.000 m length of

circular sewer of one brick (225 mm) thickness with an internal diameter of 1.000 m. (This is really a brick tube.)

3   Calculate the total volume of earth to be removed in excavating a rain-water soakaway circular on plan with a diameter of 2.100 m and a depth of 3.600 m. Allow 20% for bulking and give your answer to the nearest cubic metre.

4   A short, hollow wood column supporting a display platform has a cross-section comprising an ellipse, 175 mm by 100 mm, with a 50 mm diameter hole through it. Its length is 450 mm. If the wood has a density of 520 kg/m^3 what does the column weigh?

5   Find the volume, in cubic metres, of the concrete in a 2.000 m long concrete pipe having an internal diameter of 1.200 m and a wall thickness of 50 mm.

6   A kiln, circular on plan and 5.500 m high, has to have its curved internal surfaces lined with firebrick. It is 3.000 m in diameter. What is the area to be lined in square metres? What is the capacity of the kiln in cubic metres if the firebrick is 100 mm thick?

7   An open topped container is cylindrical in shape and constructed of sheet copper. Its diameter is 150 mm and its depth 200 mm.
(a)   Calculate the number of litres it will hold. (1 litre = 1000 cm^3)
(b)   Its interior surfaces are to be tinned. What area of tinning (in square centimetres) does this represent?

8   A tower is circular on plan and has a hemispherical domed roof. The internal diameter is 7.800 m and the wall height, from the floor to the springing of the dome, is 12 m. Calculate:
(a)   The cubic content (volume) of the tower and dome.
(b)   The surface area of the dome.

9   A sphere of lead having a diameter of 50 mm is carefully beaten out into a circular sheet one millimetre thick. What will be the diameter of this circle?

10 A hemispherical basin holds 4.500 litres of water. What is its diameter in millimetres?

11 An excavation is made in the shape of a shallow cone, the diameter across the top being 15.000 m and the depth at the centre 2.400 m. What is the resulting increase in the surface area of the ground in square metres?

12 A conical heap of sand has a diameter at the base of 2.400 m and a height at the centre of 1125 mm. How many cubic metres of sand does it contain?

13 A turret has a conical roof which is 2.250 m in diameter and 3.750 m high. Allowing 10% for waste and joints, calculate the number of square metres of sheet copper needed to cover it.

14   A balance weight is cast in the shape of a hemisphere surmounted by a cone. Its maximum diameter is 450 mm. What must be the height of its conical portion if it is to weigh 80 kg and is made of concrete of density 2240 kg/m^3?

15   The water jacket of a cast iron boiler is shaped as shown in Figure 86. It is 300 mm high. What is its capacity in litres.

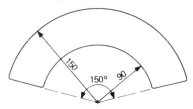

Figure 86

Revision exercise 2

1   Find the value of the following, using tables where necessary:
    (a) 23^2
    (b) 19.1^2
    (c) 7.84^2
    (d) $\left(\dfrac{1}{4}\right)^2$
    (e) $\left(\dfrac{2}{3}\right)^2$
    (f) $(1\frac{3}{4})^2$
    (g) π^2
    (h) $(8-5)^2$

2   Without using tables write down the values of the following:
    (a) $\sqrt{121}$
    (b) $\sqrt[3]{27}$
    (c) $\sqrt{\dfrac{9}{64}}$
    (d) $\sqrt{16+9}$

3   Use tables to find the following square roots:
    (a) $\sqrt{576}$
    (b) $\sqrt{9216}$
    (c) $\sqrt{196}$
    (d) $\sqrt{1225}$
    (e) $\sqrt{1521}$
    (f) $\sqrt{5184}$
    (g) $\sqrt{3025}$
    (h) $\sqrt{2401}$

4   Using the laws of indices find the values of the following:

(a) $\dfrac{3^2 \times 3^{2.4}}{3^{1.4}}$

(b) $\dfrac{1.5^3 \times 1.5^2}{1.5^{0.75} \times 1.5^{2.25}}$

5   Use logarithms to evaluate:

(a)   23 x 150 x 2.718

(b)   391.6 x 15.84 x 1.414

(c) $\dfrac{1.763}{1.032}$

(d)   1.732 x 3.142 x $(4.75)^2$

(e) $2.84^3 \times 1.96^3 \times 23.74$

(f) $\dfrac{75320}{43.5^2}$

(g) $\dfrac{324.3^2}{27.3^3}$

(h) $\sqrt{54.25 \times 3.142^2}$

(i) $\sqrt[3]{84.37 \times 23}$

6   Use logarithms to find the value of:

(a) $\dfrac{fbd^2}{6}$ when $f = 7.15$, $b = 50$, $d = 175$

(b) $\sqrt{2gh}$ when $g = 9.81$, $h = 13.56$

7   Use logarithms to evaluate the following:

(a) $3.76 \div 17.95$

(b)   0.953 x 1.846

(c) $\sqrt{63.5} \div 0.876$

(d) $\sqrt{9.81 \times 0.56}$

(e) $\dfrac{(11.31 + 14.76)}{51.73}$

(f) $\dfrac{1}{3.76} + \dfrac{1}{7.36}$

8   A window contains eight panes of glass each 450 mm square. What will the glass cost at £10.84 per square metre?

9   A room is 4.650 m long and 3.600 m wide. Its floor to ceiling height is 2.550 m. It has two door openings each 2100 mm x 900 mm and two window openings, one 2.475 m x 1.200 m and the other 1.800 m x 1.200 m. Calculate the area of wall surface for plastering to the nearest 0.1 m².

10   Figure 87 shows a building plan. Find the floor area and perimeter.

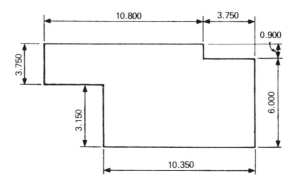

Figure 87

11   How many metres squared of brickwork are there for pointing in the end wall shown in Figure 88? The window opening is 1.350 m x 1.050 m and the door opening 2.100 m x 825 mm.

12   The floor of a cylindrical concrete storage chamber has to be hacked and rendered with a cement and sand screed. If the chamber has a diameter of 3.150 m, what area has to be treated?

13   A corner plot in the shape of a quadrant (quarter) of a circle is to be fenced. What length of fencing is required to completely enclose it, if its straight sides are 42 m long?

14   Find the length of picture rail required in a square room whose floor area is 16 m². What would it cost at 57p per metre run?

15   The plot of land shown in Figure 89 is to be split up into nine smaller plots of equal area. What will be the area of each plot to the nearest m²?

16   Find the area of each of the following circles.
(a)   Radius 1.750 m
(b)   Radius 625 mm
(c)   Diameter 2.400 m

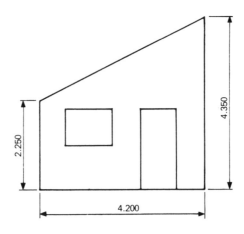

Figure 88

(d) Diameter 360 mm
(e) Circumference 24.000 m
(f) Circumference 12.750 m

17 The angle at the centre of a sector of a circle having a radius of 6.300 m is 75°. What is the area of the sector?

18 Calculate the length of the arc of a circle of diameter 250 mm which is subtended (cut off) by an angle of 36° at the centre.

19 The circumference of a circle is 17.500 m. What will be the area of a sector of the circle whose arc length is 3.500 m?

20 The shape of a sun terrace is a minor segment of a circle of radius 2.500 m. The chord length is 4.756 m. Find the area of the terrace if it subtends an angle of 144° at its centre.

21 An annulus has an internal diameter of 950 mm and an external diameter of 1.150 m. What is its area?

22 A curved portion of a concrete path has the shape of a sector of an annulus. Its outer radius is 4.500 m and its inner radius 3.000 m. The angle subtended at the centre is 50°. Find the area of path involved.

23 In a tiled floor the central portion is an ellipse having a major axis of 5.400 m and a minor axis of 3.600 m. This is to be tiled in tiles of a different colour to the remainder of the floor. Find the area of the ellipse and its perimeter.

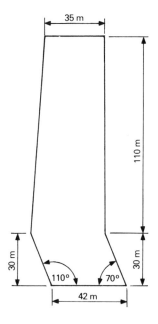

Figure 89

24   A large bay window is semi-elliptical on plan. Its width is 3.000 m and it projects 1.050 m at its centre. It has a flat roof which is to be covered with roofing felt. Find the area to be covered and the length for edge finishing.

25   The cross-section of a concrete column is a rectangle measuring 350 mm by 200 mm. If the density of the concrete is 2400 kg/m^3 and the column is 2.4 m high, find its weight.

26   A channel carrying water is semi-circular in cross-section with a diameter of 600 mm. How many litres of water will pass through it per minute if the rate of flow is 1 m per second? (Answer to nearest 10 litres.)

27   Figure 90 shows the cross-section of a plinth stone. How much will the stone weigh if it is 1 m long and the density of the stone is 1900 kg/m^3?

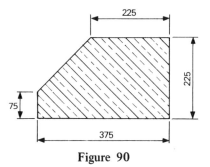

225

225

75

375

Figure 90

28   An embankment forming a protective wall between a building plot and a river is shown in Figure 91. What volume of earth in m^3 is contained in 1 kilometre of the wall?

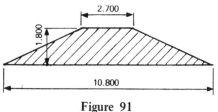

2.700

1.800

10.800

Figure 91

29   A timber beam 3.600 m long, 100 mm wide and 225 mm deep weighs 54 kg. What will be the weight of a beam made of similar timber but 4.5 m long, 150 mm wide and 300 mm deep?

30   How many cubic metres of spoil (excavated material) will result from digging a drainage trench 50 m long, 1 m wide and 1.500 m average depth, if the earth increases in bulk by 27½% when excavated?

31   Figure 92 shows the cross-section of a trench to be excavated to a length

of thirty metres. Calculate the volume of excavated earth to be carted away if it bulks by 15% when excavated.

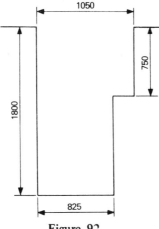

Figure 92

32 Find the volume of the space enclosed by a gable ended roof of span 4.500 m and height from eaves to ridge of 3.900 m if its dimensions on plan are 4.500 m by 9.000 m.

33 A hexagonal bar of steel is 300 mm long and the length of side of its hexagonal cross-section is 15 mm. If the density of steel is 7800 kg/m³ find the weight of the bar.

34 A garden sun lounge is octagonal on plan, the length of each of its sides being 1.200 m and the distance between parallel walls 2.897 m. Its floor to ceiling height is 2.250 m. Find:
(a) The volume of air it encloses.
(b) The area of its walls.
(c) The area of its floor.

35 Figure 93 shows the cross-section of a cutting for a service road to an estate. The length of the road is 15 m. Find the volume of earth removed to form the cutting.

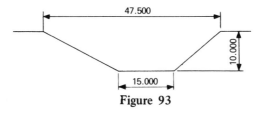

Figure 93

36 Calculate the capacity in litres of a cylindrical tank having a diameter of 2.250 m and a depth of 1.850 m (1 litre = 1 dm³).

37   A circular tower of height 7.200 m and diameter 4.500 m is to have its walls rendered. Find the area for rendering (externally) in m².

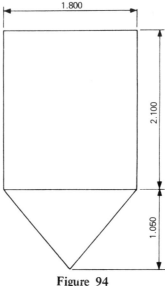

Figure 94

38   A 300 mm length of stoneware pipe has an internal diameter of 100 mm and a wall thickness of 20 mm. Calculate its weight if its density is 2243 kg/m³.

39   A domed roof is in the shape of a hemisphere. Its diameter is 7.500 m internally. Find the volume it encloses and its surface area.

40   The spherical float of a ball valve is of copper having a density of 8900 kg/m³. It has an external diameter of 125 mm and the thickness of the metal is 0.5 mm. Calculate its weight.

41   A hopper is square on plan and has the shape shown in Figure 94, the lower portion being an inverted pyramid. What is its capacity?

42   A pyramidal roof of height 3.600 m has a hexagonal plan, the length of side of the hexagon being 1.500 m. Find its surface area.

43   Find the volume and surface area of a cone with a base diameter of 100 mm and height 150 mm.

44   A north light roof truss has its steep slope pitched at 60° and its other side at 30°. Its span is 4.800 m. Calculate the two rafter lengths.

45   The rafter length of a lean-to roof pitched at 45° is 4.200 m. What is the height of the roof?

9 Formulae

The use of symbols and letters is of great importance in mathematics. You are already familiar with a number of the more common symbols used, for example:

+ means *plus* or *add*
- means *minus* or *subtract*
x means *multiply by*
÷ means *divide by*
= means *is equal to*

There are many others in use and here are a few more quite useful mathematical symbols:

≏ or ≐ means *is approximately equal to*
≡ means *is identical with*
≠ means *is not equal to*
< means *is less than*
> means *is greater than*

and, of course, π

Symbols and letters also have a special meaning in *algebra*, which is really a branch of mathematics, where *letters* such as a, b, x and y are used instead of numbers, until their real numerical value is known (this will be explained later in more detail).

A *formula* is a way of expressing the value of a quantity in terms of other values directly associated with it.

An example we have already met is the formula $A = l^2$, which you will remember gives the area, A, of a square in terms of the length of its side, l.

Another is $A = l \times b$ which gives the area, A, of a rectangle in terms of its length, l, and breadth, b.

These are very simple formulae. An understanding of them, and others, will help a great deal in the solution of many problems.

In Chapter 6 we saw that the circumference of a circle may be found by multiplying twice the radius by π. As a formula this reads: $C = 2\pi R$.

Similarly, the area of a circle, which we saw to be the radius squared then multiplied by π, is expressed as: $A = \pi R^2$.

Notice that there are no multiplication signs used. When two symbols are written side by side without a sign between them it is understood that they are to be multiplied.

At this stage in your work you may be asked to use formulae in either of the following ways:

1   You may be asked to solve a problem and be expected to know the formula involved.

2   You may be given a formula and asked to find its value from given information.

So you see it is necessary for you to *learn* and *remember* a number of the more useful formulae encountered in your work and also to be able to work them out.

Example 1

Calculate the area of a circle which has a radius of 3.500 m. We know from previous work that: area = π x *radius* x *radius* $(A = \pi R^2)$.

Now *substitute* the values given, that is:

$R = 3\frac{1}{2}$

$\pi = 3\frac{1}{7}$

$A = 3\frac{1}{7} \times 3\frac{1}{2} \times 3\frac{1}{2}$

$= \frac{\overset{11}{\cancel{22}}}{7} \times \frac{7}{\cancel{2}} \times \frac{7}{2}$

$= \frac{77}{2}$

$= 38\frac{1}{2}$ m^2

$= 38.5$ m^2

There is no reason why vulgar fractions should not be used in a metric system if they give an easy method of solution in this way.

Example 2

The density of a substance, D, is related to its weight, W, and its volume, V, by the following formula:

$$D = \frac{W}{V}$$

What is the density of a block of stone weighing 200 kg with a volume of one tenth of a cubic metre?

Having been given the formula, first write it down:

$$D = \frac{W}{V}$$

Now look for the correct values to substitute:

$W = 200$ kg
$V = 0.1$ m^3

$D = \frac{200}{0.1}$ kg/m^3
  $= 2000$ kg/m^3

A word of warning: Make sure that your figures are in the *correct units* before evaluating the formula.

Example 3

Find the value of A in the following formula correct to three decimal places.

$$A = \frac{BH}{2} \text{ where } B = 1.450 \text{ m and } H = 450 \text{ mm}$$

Look carefully and you will probably recognize this formula. It gives the area, A, of a triangle in terms of its base, B, and height, H (see Chapter 6). But notice that B is given in metres, whereas H is given in millimetres. It is necessary to express both in the *same units* before substituting them in the formula. Thus:

$$A = \frac{BH}{2}$$
$$= \frac{1.450 \times 0.450}{2}\,\text{m}^2$$
$$= 0.326 \text{ m}^2$$

$$\begin{array}{r} 1.45 \\ .45 \\ \hline .580 \\ .0725 \\ \hline 2)\overline{.6525} \\ .32625 \end{array}$$

It is not always necessary to recognize a formula in order to evaluate it, provided all the required information is given. It may look quite complicated but will often work out quite easily after the correct values have been substituted.

Example 4

Find the value of M in the formula

$$M = \frac{WL}{4} - \frac{wL}{8} \text{ where } W = 5, L = 20, \text{ and } w = 2$$

Sufficient information is given to substitute the values required to find M.

We must assume that the units are correct.

$$M = \frac{WL}{4} - \frac{wL}{8}$$

$$= \frac{5 \times \overset{5}{\cancel{20}}}{\cancel{4}} - \frac{2 \times \overset{5}{\cancel{20}}}{\underset{4}{\cancel{8}}}$$

$$= 25 - 5$$

$$= 20$$

Since no units are given in this case, the answer is a number.

Construction of formulae

So far we have used established formulae, but the following examples will show you how to *construct* your own formulae.

Example 5

A salesman is paid x pence per hour for a 48 hour week plus 50 pence commission on every sale made. Construct a formula for his weekly wage, using W to represent the wage and N to represent the number of sales made.

This is not as complicated as you may at first think if you analyse if carefully.

If he works 48 hours for x pence per hour this will bring him 48x pence. In addition N sales at 50 pence per sale will bring him 50 N *pence*. So his total earnings will be the sum of these two amounts. That is:

$W = 48x + 50\,N$ in pence, or
$W = 0.48x + 0.50\,N$ in pounds

Example 6

Using the formula constructed in Example 5, obtain the wage of the salesman in a week when his hourly rate is 180 pence and he makes 50 sales.

Using the formula to give pounds:

$W = 0.48x + 0.50\,N$
  Substituting 180 for x and 50 for N:
$W = £(0.48 \times 1.80) + (0.50 \times 50)$
   $= £86.40 + £25$
   $= £111.40$

Example 7

In order to convert a temperature in degrees Celsius to degrees Fahrenheit you must divide it by 5, multiply by 9 and then add 32 to the result. Using C for the Celsius temperature and F for the corresponding Fahrenheit temperature, express this method of conversion as a formula. Use this formula to find the Fahrenheit temperature corresponding to 15°C.

Carry out the instructions given:

$$C \div 5 = \frac{C}{5}$$

Now, multiply by 9, giving:

$$\frac{C}{5} \times 9 = \frac{9}{5}C$$

Next, add 32, giving:

$$\frac{9}{5}C + 32$$

So the formula is:

$$F = \frac{9}{5}C + 32$$

Substituting 15 for C:

$$F = \left(\frac{9}{\cancel{5}_1}\right) \times 1\cancel{5}^3 + 32$$
$$= 27 + 32$$
$$= 59°F$$

Example 8

A precast concrete coping has the cross-section shown in Figure 95. Construct a formula for the volume in cubic metres, of a length L metres of the coping and use it to find the volume of a 12 m length, when b = 300 mm, s = 75 mm and t = 100 mm. (Ignore the drips.)

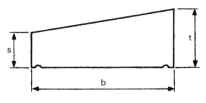

Figure 95

A length of this coping will, of course, form a *prism* (see Chapter 6), and its volume will be the area of its cross-section (a trapezium) multiplied by its length.

Now, the area of a trapezium is *half the sum of the parallel sides times the perpendicular distance between them* (see Chapter 6).

In this case, then:

$$\text{area of cross-section} = \frac{s+t}{2} \times b$$

which is better written $b\dfrac{(s+t)}{2}$

Now, multiply by the length, L, giving:

$$Lb\frac{(s+t)}{2}$$

or, using V for volume:

$$V = Lb\frac{(s+t)}{2}$$

Now, substituting the given values:

$$V = 12 \times 0.300\left(\frac{0.075 + 0.100}{2}\right)$$
$$= 12 \times 0.3 \times 0.0875$$
$$= 0.315 \text{ m}^3$$

Note that since the volume is required in m³ the dimensions must all be expressed in m when they are substituted in the formula.

Example 9

Construct a formula for the area, A, of the flat plate shown in Figure 96. Use the symbols given as dimensions.

This is quite an easy one to plan out. The plate consists of a rectangle, a circle (two semi-circles) and a square hole. We must add together the areas of the rectangle and the circle and subtract the square.

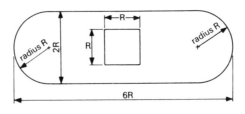

Figure 96

Area of rectangle = length x breadth
$$= 4R \times 2R$$
$$= 8R^2$$

Area of circle $= \pi R^2$

Area of square $= R^2$

Add together the rectangle and the circle:

$8R^2 + \pi R^2$

Subtract the square:

$8R^2 + \pi R^2 - R^2$

Now, since all the terms in this expression contain R^2 the expression may be factorized, which means taking the *common factor* (R^2) and writing it outside of a bracket, the bracket containing the other factor of each term. That is:

Area $= R^2 (8 + \pi - 1)$

(The last term was, of course, $1R^2$ although the 1 was not written.)

Substituting 3.14 for π and writing A for area we have:

$$A = R^2 (8 + 3.14 - 1)$$
$$= 10.14 R^2$$

Example 10

A rectangular room has a window on one of its long walls and a window on one of its short walls. The width of each window opening is one half of the wall containing it and both windows are two thirds of the height of the room.

(a) Using *1, b,* and *h* for the length, breadth and height of the room, construct a formula expressing the total window area in terms of the room's dimensions.

(b) Express the window area as a fraction of the gross wall area, A. A diagram of the four walls developed (opened out) will help. (See Figure 97.)

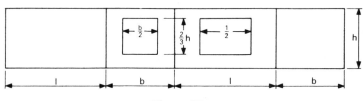

Figure 97

(a) Window area $= \frac{2}{3}h \times \frac{1}{2}b + \frac{2}{3}h \times \frac{1}{2}l$

Cancel the 2s and $h/3$ is a common factor. Writing W for window area:

$$W = \frac{h}{3}(b + l)$$

(b)   The developed rectangle has a length of $2b + 2l$ or, taking out the common factor, $2(b + l)$. Multiplying this by the height we have the area:

$$A = 2(b + l) \times h$$
$$= 2h(b + l)$$

The required fraction is:

$$\frac{\text{window area}}{\text{wall area}} = \frac{W}{A}$$

$$\frac{W}{A} = \frac{\frac{h}{3}(b+l)}{2h\,(b+l)}$$

Cancel the hs and the $(b + l)$s leaving:

$$\frac{W}{A} = \frac{1/3}{2}$$
$$= \tfrac{1}{6}$$

Example 11

In a building firm the technicians receive, on average, two thirds of the pay of the administrative staff and the craftsmen receive two thirds of the pay of the technicians. There are twice as many technicians as administrative staff and the craftsmen outnumber the technicians by twenty to one. If the average weekly wage of each administrator is £L and there are n of them, construct a formula for the firms weekly wage bill £W in terms of L and n.

Starting with the administrative staff:

Number employed $= n$
Average wage $= £L$
Weekly bill $= £nL$

Now the technicians:

Number employed $= 2n$
Average wage $= 2/3L$
Weekly bill $= £\, 2n \times 2/3L$
                   $= £\dfrac{4nL}{3}$

Finally, the craftsmen:

Number employed $= 20 \times 2n$
$= 40n$

Average wage $= £\dfrac{2}{3} \times \dfrac{2}{3}L$

$= £\dfrac{4}{9}L$

Weekly bill $= £40n \times \dfrac{4}{9}L$

$= \dfrac{£160nL}{9}$

Total weekly bill $= £nL + \dfrac{4nL}{3} + \dfrac{160nL}{9}$

Taking out the common factor, nL, we have:

$W = £nL \left(1 + \dfrac{4}{3} + \dfrac{160}{9} \right)$

$= £nL \ \dfrac{9 + 12 + 160}{9}$ (common denominator)

$= £nL \times \dfrac{181}{9}$

$= \dfrac{£181nL}{9}$

Evaluation of a formula

The process of evaluation may be divided into three parts;

1 Substitution – replacing the symbols by their respective values.
2 Simplification – collection of the values into groups.
3 Evaluation – the final calculation, often conveniently carried out with the aid of logarithms.

The following examples illustrate this method.

Example 12

The formula $S = 2\pi r(h + r)$ gives the total surface area of a closed cylinder of radius r and height h. Find the surface area of such a cylinder having a radius of 725 mm and a height of 1750 mm.

Substitution: $S = 2\pi r(h + r)$
$= 2 \times 3.142 \times 725(1750 + 725) \ \text{mm}^2$
Simplification: $= 2 \times 3.142 \times 725 \times 2475 \ \text{mm}^2$
Evaluation (by logs) $= 11270000 \ \text{mm}^2$
$S = 11.27 \ \text{m}^2$

No.	Log
2	0.3010
3.142	0.4972
725	2.8603
2475	3.3936
11270000	7.0521

Note: The result is a very large number when expressed in mm² but is easily converted to m² by dividing by a million (that is, by moving the decimal point six places to the left).

Example 13

The modulus of elasticity of a structural material is given by the formula $E = Wl \div Ax$, where A is the cross-sectional area and x the extension of a length l of the material produced by a load W.

Find the modulus of elasticity of the steel in a bar of cross-sectional area 1100 mm² which is stretched by 0.875 mm from its original length of 3.000 m by a load of 66 kN.

Substitution: $E = \dfrac{Wl}{Ax}$

$$= \frac{66 \times 3000}{1100 \times 0.875} \text{kN/mm}^2$$

Simplification: $= \dfrac{180}{0.875}$

Evaluation: $E = 205.7 \text{ kN/mm}^2$

No.	Log
180	2.2553
0.875	1.9420
205.7	2.3133

Note: To obtain the correct units substitute the units *only* in the formula as follows:

$E = \dfrac{Wl}{Ax}$

$= \dfrac{\text{kN} \times \text{mm}}{\text{mm}^2 \times \text{mm}}$

$= \dfrac{\text{kN}}{\text{mm}^2}$ (kilonewtons per square millimetre)

Example 14

The volume of a sphere is given by the formula $V = \frac{4}{3}\pi R^3$, where R is the radius. Find the volume of a sphere having a radius of 1.75 cm.

Substitution: $V = \frac{4}{3}\pi R^3$

Simplification: $= \frac{4}{3} \times 3.142 \times 1.75^3$ cm^3

Evaluation: $V = 22.45$ cm^3

No.	Log
1.75	0.2430
	3
1.75^3	0.7290
4	0.6021
3.142	0.4972
	1.8283
3	0.4771
22.45	1.3512

Example 15

The formula $M = f\dfrac{bd^2}{6}$ gives the moment of resistance of a timber beam whose cross-sectional dimensions are breadth b and depth d. f represents the safe working stress for the timber. Find the moment of resistance of a beam of breadth 63 mm and depth 180 mm made of timber which may be safely stressed up to 7.5 N/mm^2 (newtons per square millimetre).

Substitution: $M = f\dfrac{bd^2}{6}$

Simplification: $= \dfrac{7.5 \times 63 \times 180^2}{6}$ N mm

Evaluation: $M = 2,552,000$ N mm

No.	Log
180	2.2553
180^2	4.5106
63	1.7993
7.5	0.8751
	7.1850
6	0.7782
2552	6.4068

The units are found in the same manner as used in Example 13. See if you can do this as an exercise. Note that you should obtain N mm (newton milli-metres, a moment being a force times a distance).

Example 16

The safe load which can be carried by a steel plate weakened by rivet holes is given by the formula, $P = f(b - nd)t$, where:

t = plate thickness
b = plate breadth
d = diameter of rivet holes
n = number of rivet holes
f = safe working stress in steel

Find the safe load for such a plate 112 mm broad and 15 mm thick having two 19 mm diameter rivet holes. The safe working stress for the steel is 150 N/mm² (newtons per square millimetre).

Substitution: $P = f(b - nd)t$
$\qquad\qquad = 150 \, (112 - 2 \times 19) \, 15$
Simplification: $= 150 \, (112 - 38) \, 15$
$\qquad\qquad = 150 \times 74 \times 15$
Evaluation: $P = 166500 \, \text{N}$
$\qquad\qquad = 166.5 \, \text{kN}$

Here the answer in newtons (N) has been divided by 1000 to give kilonewtons (kN).

Note: See Chapter 11 for an explanation of the newton (N) as a unit of force.

Exercise 15

1   Construct a formula giving the area, A, of each of the following figures shown in Figure 98 (a), (b) and (c) in terms of the given dimensions.

2   In a hall the stage is two thirds the width of the hall and its depth (back to front) is one sixth the length of the remainder of the hall. Seating accom-modation for the audience occupies three quarters of the floor space excluding the stage.
    (a)   Construct a formula giving the area A of the stage in terms of the length l (excluding the stage) and the breadth b of the hall.
    (b)   Express the stage area as a fraction of the seating area.

3   On a building site the craftsman's weekly pay is three quarters of the fore-man's. Apprentices get an average of half of the craftsman's pay and

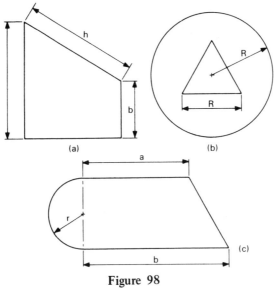

Figure 98

labourers get three quarters of the craftsman's.

The one foreman on the site has control of four times as many crafts-men as apprentices and half as many labourers as craftsmen. Construct a formula to give the total weekly wage bill, W, for the site in terms of the number of craftsmen, n and the craftsman's weekly wage £x.

4   The surface area of a closed cylinder of radius R is given by the formula, $S = 2\pi R(H + R)$, where H is the height of the cylinder. Find the surface area of such a cylinder having a radius of 0.900 m and a height of 1.200 m.

5   The following formula is used to obtain the value of the moment of resist-ance of a timber beam having a rectangular cross-section, b and d being the breadth and depth of the beam and f the value of the safe working stress for the timber.

$$M = \frac{f b d^2}{6}$$

When rearranged to the form

$$d = \sqrt{\frac{6M}{fb}}$$

it may be used to find the depth of a beam. Find the value of d when $M = 2,552,000$, $f = 7.5$ and $b = 63$.

6   Write down the S formula for the area of a triangle and use it to find the area of a triangle whose sides are 3.800 metres, 4.25 metres and 2.600 metres.

7   The thickness of metal in a cast iron pipe is one fifth of the square root of the pipe's diameter. Write this down as a formula using t for thickness, and d for diameter. Find the thickness of metal in such a pipe whose diameter is 225 mm.

8   The formula for the volume V of a sphere of radius r is, $V = \frac{4}{3}\pi r^3$.

   (a)   Write down a formula for the weight of a sphere composed of material of density W kg/m³.

   (b)   Calculate the weight of a sphere of cast iron of radius 50 mm if the density of cast iron is 7208 kg/m³.

9   Write down the formula for the volume of a cylinder using V for volume and r for radius and h for height. Now calculate the total capacity in litres of a cylindrical water storage tank of height 4.200 m and diameter 7.500 m. (1 m³ = 1000 litres.)

10   Two cylindrical stone columns each 1125 mm diameter and 6 m high are to be demolished and carted away as rubble. The contractors price for breaking up and removal is £28.50 per tonne. The stone weighs 2240 kg/m³. Write a formula giving the cost of the work in terms of radius, r; height, h; density, w; unit cost, p; and use it to calculate the cost of the job to the nearest £1. (1 tonne = 1000 kg.)

11   The stress, f N/mm², in a tie rod is given by the formula $f = \frac{W}{A}$, where W is the applied load and A is the cross-sectional area of the rod. Find the breaking stress in a steel rod of diameter 30 mm which breaks under a load of 600 kN.

12   A tank, circular on plan, is to have an internal diameter of 1500 mm. What must be its depth for it to have a capacity of 4550 litres? (1 m³ = 1000 litres.)

10 Graphs

A graph provides a means whereby the relationship between two quantities can be represented in pictorial form. A great deal of information can be obtained from a graph, and there are many different types. In this chapter I shall give only an introduction to graphical work and apply this knowledge to the type of graph associated with statistics and experimental work.

Plotting a graph

Constructing a graph is usually referred to as *plotting*, and a graph consists of a number of points plotted in relation to two fixed lines set at right angles to each other, known as *axes* (the plural of *axis*). These points are usually joined to form a line, which can then be 'read'.

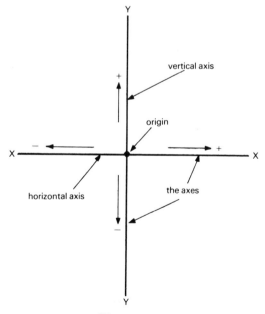

Figure 99

Figure 99 is an illustration of the complete graph *axes.* Note the zero position called the *origin.*

Measurements taken upwards or forwards (to the right) are positive (or +) measurements.

Measurements taken downwards or backwards (to the left) will be negative (or -) measurements.

In problems where *only positive* or + measurements are concerned, it is only necessary to use that part of the axes as shown in Figure 100.

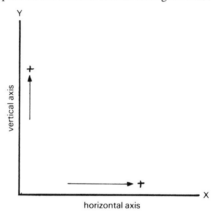

Figure 100

In practical problems, the axes must always be *named*, otherwise the graph has no meaning; but for general purposes it is usual to call the horizontal axis x, and the vertical axis y.

Before beginning to plot your points on a graph, it is first necessary to draw the axes required and then to divide them into the required numbers of parts.

This should be done very carefully and neatly, and with a view to obtaining the largest graph line possible in the space available. Figures 101 and 102 illustrate a very common error.

Points may now be plotted on the graph by fixing the position of each one relative to the appropriate axis.

Example 1

Plot the following points on the same graph:
(a)   (3, 5)
(b)   (- 2, 4)
(c)   (5, -5)
(d)   (-3, -3)

Note: In this method of notation the *x value is always given first* followed by the *y* value.

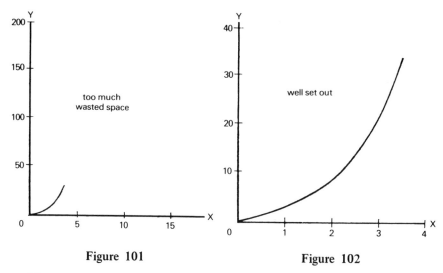

Figure 101   Figure 102

There are negative values involved so the full graph axes will be required.

1   Set up the axes,
2   Subdivide them carefully.
3   Plot each point separately. (See Figure 103.)

Types of graphs

Pictorial graphs

Some graphs have a purely pictorial value for example temperature charts, sales graphs, etc.; they indicate a *trend* and the shape or inclination of the line (which is usually irregular) does not normally follow any mathematical law or rule.

In such graphs, the points are best joined by straight lines and it is not possible to read off intermediate values on the line of the graph.

Example 2

Suppose that the following table represents an analysis of your homework marks for craft calculations for the autumn term:

Mark (max. 10)	3	4	4	5	4	0 abs.	4	6	6	8	7	8
Week	1	2	3	4	5	6	7	8	9	10	11	12

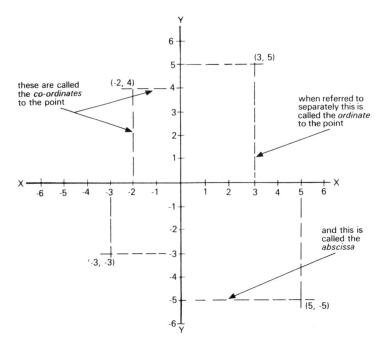

Figure 103

Plot a graph showing these results.

Note:

1   Notice the method of tabulating the results in the form of a *graph table*. Each pair of figures indicates a single point on the graph.

2   There are no negative values so the axes may be set out and divided as shown in Figure 104.

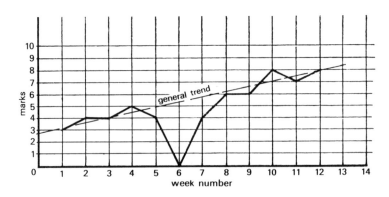

Figure 104

3   Try to make your graph look *neat* and *professional*. Mark the points with a small dot or compass prick, *not* a hefty black cross.

4   Do not forget to name each axis.

5   Although the divisions along each axis must be equal, there is no necessity to make the divisions on the horizontal axis equal to the divisions on the vertical axis.

It is obvious that except for the week you were absent your maths marks have shown a fairly steady improvement.

Straight line graphs

Where it is *known* that the graph is a *straight line*, it is only really necessary to plot three points (the third point being included as a check on the accuracy of the other two).

Exercise 16

1   The following table shows the relationship between the area of a circle and its radius.

   (a)   Construct a graph showing this relationship.
   (b)   Find from it the area of a circle whose radius is 55 mm.
   (c)   Check the answer to (b) by calculation.

Radius (cm)	1	2	3	4	5	6	7
Area (cm²)	3.142	12.57	28.28	50.27	78.55	113.11	153.96

Note: The graph should be drawn as a smooth curve. If it is really accurately drawn you will be able to read off the area of any circle from 1 cm to 7 cm radius at a glance.

2   In order to compare their rates of absorption, three different types of brick were immersed in water and weighed periodically. The table below shows the results obtained. Plot a graph of each set of results, all on the same pair of axes.

Absorbed water in kg/m³ of brick	*Immersion time in minutes*								
	1	2	3	5	10	20	30	60	120
Staffordshire blue brick	1.60	2.40	3.20	4.00	4.20	4.32	4.48	4.80	5.60
Fletton pressed brick	16.0	32.0	64.0	84.1	100.1	116.2	128.2	124.1	125
Red rubber brick	96.1	128.2	144.0	152.0	168.2	180.0	184.0	188.0	192.0

3   It is known that the graph of the relationship between the Fahrenheit and Celsius scale on a thermometer is a straight line.

(a)   Plot a graph showing this relationship using the temperatures given in the table.

(b)   Find from the graph the Celsius reading corresponding to $99°$ Fahrenheit.

(c)   Extend the graph to find the Fahrenheit reading corresponding to $110°$ Celsius.

Celsius	$0°$	$50°$	$100°$
Fahrenheit	$32°$	$122°$	$212°$

4   Plot the following conversion table in the form of a graph.

Number of Petrograd standards of timber	1	5	10
Equivalent volume in m³	4.67	23.35	46.70

Find from your graph:

(a)   The number of standards of timber contained in 14 m³ of timber.

(b)   The number of standards of timber contained in 800 m run of 75 mm x 150 mm. (Work out the volume of timber first, then read from the graph.)

(c)   The number of m³ of timber in 5.25 standards.

5 In order to compare the rate at which heat is lost through equal thicknesses of *wet* concrete and *dry* concrete, the following experimental results were obtained.

Wet concrete

Time in minutes	1	2	3	4	5	10	20	30	40
Temperature °C	60°	55°	58°	52°	35°	25°	15°	7°	5°

Dry concrete

Time in minutes	1	2	3	4	5	10	20	30	40
Temperature °C	65°	57°	52°	48°	45°	36°	28°	18°	15°

(a) Illustrate the comparison by plotting both sets of results on the same axes. (Watch for experimental errors.)

(b) Which of the two materials has the better *heat insulative* properties?

6 Draw up a graph table showing the areas of squares with sides of length 2, 4, 6, 8 and 10 cm respectively. Plot the graph and use it to find:

(a) The area of a square of side 4.75 cm.

(b) The length of side of a square of area 56 cm².

Experimental graphs

You may be asked to plot a graph from a series of results from an experiment, either carried out by yourself, or offered to you in the form of a table of values.

Much more information may be deduced from most graphs of this nature.

1 The graph may consist of a straight line or regular curve, indicating a definite mathematical relationship between the quantities. (It is shown in Chapter 9 how this relationship can be expressed in the form of an equation or formula.)

2 It is important to note that because of *experimental errors* some points on the graph may deviate from the position of the general line or curve. The line should normally be drawn to pass through the majority of the points, or to be an *average line* between them.

The given points having been plotted and the graph line drawn, it is possible to obtain values *other than those of* the plotted points by reading off intermediate values from the curve itself. This may be most useful in applied work, but note that it is possible only where the graph consists of a regular line or curve, and it is very necessary to draw the curve carefully and accurately.

Choice of scales is a particular pitfall for the learner. It is generally correct to say use the largest possible scale for the size of paper you have, but always allow a margin outside the axes for adequate labelling. Clear labelling of what is represented on an axis, and in what units, is an essential part of good presentation.

Example 3

The density of steel is 7800 kg/m^3. Plot a graph to show the variation in weight of various lengths (up to 6.000 m) of a steel joist having a cross-sectional area of 35 cm^2. From the graph read off the weight of 3.450 m of the joist and the length which will weigh 120 kgf. This will be a straight line graph because weight will vary directly with length. It will pass through the origin because no length has no weight.

Before the graph can be drawn a calculation is necessary. Since the area is 35 cm^2 we can find the volume of a 6.000 m length. That is:

$$\text{Volume} = \frac{35}{100^2} \times 6 \text{ m}^3$$
$$= 0.021 \text{ m}^3$$
$$\text{Since density} = \frac{\text{mass}}{\text{volume}}$$
$$\text{mass} = \text{volume} \times \text{density}$$

You will realize that, as previously explained, we are using mass units to represent weight. Therefore:

Weight of 6.000 m length = 0.021 × 7800 kgf
= 163.9 kgf

No.	Log
35	1.5441
6	0.7782
	2.3223
100^2	4.0000
2100	2.3223
7800	3.8921
163.9	2.2144

The graph can now be drawn (see Figure 105). You will see that the length is represented along the *x* axis and the weight along the *y* axis. It is conventional to plot the dependant variable on the *y* axis and in this case weight depends on length.

The values asked for are shown on the graph. Notice the labelling of the *y* axis is done so that one reads upwards with the head turned to the left. The reading off of intermediate values is called *interpolation*. Only by using the largest possible scales will accuracy be obtained.

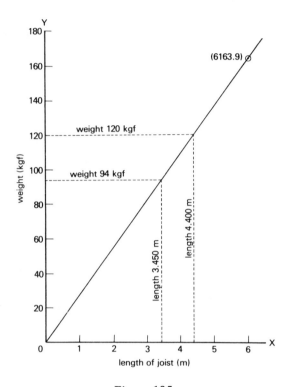

Figure 105

Example 4

Plot a graph to convert inches to millimetres, up to a maximum length of one foot, given that one inch is equal to 25.4 mm.

This is an exact conversion figure. You are probably aware that for many practical purposes it is 'rounded off' to 25 mm. The resulting error is very small and within reasonable tolerances for many practical purposes in building. Again, the graph is a straight line since the two variables, inches and millimetres vary directly with one another. It will of course pass through the origin.

To establish the line calculate the value of 1 ft in mm.

1 ft = 12 x 25.4 mm
$\quad$ = 304.8 mm

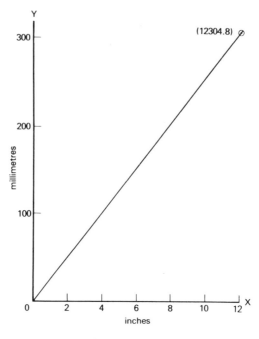

Figure 106

Now draw the graph (see Figure 106). Since we are finding the number of millimetres this dependant variable will go on the *y* axis, as before.

From the graph you can read off conversions very quickly with fair accuracy. This is a useful method of approximate conversion from imperial to metric units but you will only obtain accurate results by using large scales.

Example 5

The table below shows values of efforts required to lift various loads using a pulley system. These values were obtained by experiment and are subject, therefore, to small errors. Plot the points and draw an average line through them. From the graph find by interpolation the effort required to lift a load of mass 76 kg, and the load it is possible to lift by applying an effort of 112N.

Effort (N)	50	67	83	100	120	132	148	168	184	200
Load (kg)	10	20	30	40	50	60	70	80	90	100

The graph is shown in Figure 107.

The effort is the dependant variable this time (its value depends on the size of the load) so it is plotted on the *y* (vertical) axis.

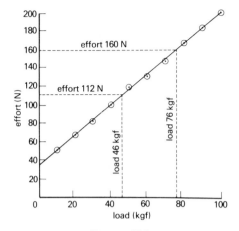

Figure 107

When you have chosen a suitable scale and plotted the points from the table of values you must 'smooth out' the experimental errors by drawing a 'fair' or average line through the points. In this case you see that the plotted points correspond very closely to a straight line and the average line can be drawn fairly easily. Experimental error is often greater than this. Having drawn the line the required values may be read off and they are shown on the graph.

Take note of the fact that the line does not pass through the origin. It cuts the *y* axis at a value of 34N. This is because the pulley system would require a force of 34N to move it when free from load.

Example 6

A timber beam having cross-sectional dimensions of 75 mm breadth and 175 mm depth can support the distributed loads shown in the table over the given spans.

Plot a graph of these values and from it, find the safe distributed load the beam will carry over a span of 2.700 m and the maximum span over which it may be used to support a distributed load of 600 kg.

Span (m)	0.5	1	1.5	2	2.5	3	3.5	4
Distributed load (kN)	43	21.5	14.33	10.75	8.16	7.16	6.14	5.38

You will notice (see Figure 108) that as the span increases the safe load decreases. Plot the points carefully and you will find that they form a curve. The line joining them should be a smooth curve with no sudden changes in direction.

The required values are obtained by interpolation as shown on the graph.

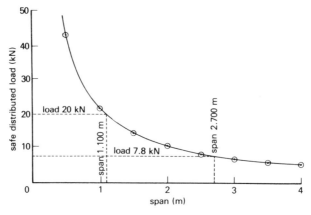

Figure 108

When drawing a curve like this it is best to turn the paper so that the swing of your wrist goes with the curve. Don't hold your pencil too tightly and try to keep your eye ahead of the pencil point. With a bit of practice quite accurate curves can be drawn freehand.

Example 7

The shank diameters of wood screws between screw gauge no. 1 and screw gauge no. 10 vary directly as the gauge number. Given that the shanks diameter of a no. 1 screw is 1.65 mm and that of a no. 10 screw 4.80 mm, construct a graph to give, by interpolation, the shank diameters of all the intermediate gauge numbers (see Figure 109).

In this case the shank diameter is dependant on the gauge number, so we plot the diameter on the y axis. We need a scale enabling us to read to no. 10 on the

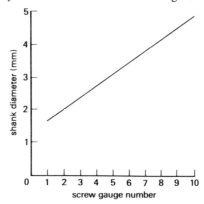

Figure 109

x axis and to 5 mm on the *y* axis. The two given points can then be plotted and joined by a straight line. All the intermediate gauge numbers may now be read off as shank diameters. For example:

Shank diameter of no. 4 screw = 2.7 mm
Shank diameter of no. 8 screw = 4.1 mm

You can also use this graph in reverse. Suppose you wanted to know the best screw size for a 4.50 mm diameter hole. The graph indicates that a no. 9 screw has a 4.45 mm shank which would just have clearance in a 4.50 mm hole.

Example 8

The following table gives the weights per metre length of steel reinforcing bar for concrete construction. Plot a graph of the given values. What would be the weight of one metre of 5 mm diameter bar and the diameter of bar having a weight of 2.5 kgf/m?

Diameter (mm)	6	10	14	18	22	25
Weight per metre (kgf)	0.225	0.62	1.23	2.02	3.06	3.96

This graph is another curve (see Figure 110). Again it is a smooth curve and should be drawn carefully in the way previously described. The values asked for are shown on the graph.

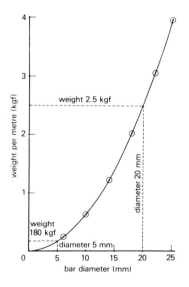

Figure 110

To summarize the work of this chapter, you should:

1   Choose an easy, large scale.
2   Use the y axis for the dependant variable.
3   Label the axes clearly.
4   Plot the points carefully.
5   Draw the curve carefully, making it an average line if the points are known to be subject to error.
6   Read off required information accurately.

Exercise 17

1   The pressures at different depths in a liquid measured experimentally are given in the table below. Plot a graph of these values and from it find:
(a)   The pressure at a depth of 275 mm.
(b)   The depth at which the pressure is 185 kN/m^2.

Depth below surface (mm)	100	200	300	350
Pressure (kN/m^2)	157	312	480	560

2   Given that the density of water is 1000 kg/m^3 and that its imperial equivalent is 62.5 lb/ft^3, construct a graph from which you can convert from lb/ft^3 to kg/m^3 up to a maximum of 3000 kg/m^3. From the graph convert the following densities to kg/m^3:
(a)   Concrete, 144 lb/ft^3.
(b)   Pitchpine, 42 lb/ft^3.
(c)   Common brickwork, 125 lb/ft^3.

3   The rate at which a tank is being filled from a supply pipe is shown in the table below. Plot a graph of these values and use it to find:
(a)   How long from the start of filling will the tank hold 250,000 litres?
(b)   How much water was in the tank before filling was started?
(c)   How much water was in the tank 18 minutes after filling started?

Time from start of filling (min.)	5	10	15	20	25	30
Litres in tank	57000	90000	123000	156000	189000	222000

4   Plot a graph showing the relationship between the area of a square and the length of its side up to a value of 10 metres. From your graph find the floor area of a square room of side 3.750 m. Also find from your graph the value of $\sqrt{75}$.

5 A saturated brick was weighed periodically as it dried out and the following values obtained.

Time (days)	0	2	4	6	8	10	12	14	16	18	20
Water in brick (grams)	270	180	115	80	60	48	37	28	22	17	14

Plot the graph of these values and from it find the weight of water lost in the first week and the time taken for the brick to lose two thirds of the absorbed water.

6 A set of values of loads lifted with a Weston pulley block and the effort required for each load is shown below. Plot these values and draw the average line through them. From your graph find the effort required to lift a load of mass 20 kg and the load which could be lifted by an effort of 30 N.

Load (kg)	4.5	9	13.5	18	22.5	27
Effort (N)	14.7	21.4	28.5	35.2	42.3	48.8

7 The safe distributed loads for various spans for a certain steel beam are shown below. Plot these values and draw the curve connecting them. From the graph determine:
(a) The safe load over a span of 4.800 m.
(b) The maximum span for a load of 320 kN.

Span (m)	2.4	3.0	4.2	5.4	6.0	6.6
Safeload (kN)	447	373	267	207	187	170

8 If 1 kg = 2.205 lb draw a graph to convert masses of up to 12 lb to kg correct to the nearest 0.05 kg. From it convert the following:
(a) 5 lb 8 oz ⎫
    7 lb 4 oz ⎬ to kg
    9 lb 12 oz ⎭
(b) 3.45 kg ⎫
    1.05 kg ⎬ to lb (nearest 0.1)
    4.75 kg ⎭

9 The following values of maximum spans and safe distributed loads are for a 150 mm thick precast concrete slab floor.

Span (m)	4.2	4.65	5.4	6.3	7.5
Load (kN/m²)	10	7.5	5	3	1.5

Plot a graph showing this relationship and from it determine:
(a) The maximum span for a floor load of 4 kN/m².
(b) The safe distributed load to be carried over a span of 4.500 m.

11 Force, motion and equilibrium

You have probably dealt with problems involving *force* in your craft science lectures. The forces most commonly met are due to gravity, or *weights*. Thus all *loads* constitute forces. Unlike many of the quantities we have been dealing with you will realize that the *direction* of a force is as important as its size or magnitude. Clearly an *upward* force has an entirely different effect to a *downward* force on the same object, even though both may be of equal magnitude. Quantities which have direction in this way are known as vector quantities; *velocity* is another, and *acceleration* a third vector quantity.

A force may be a push or a pull; for instance, a load may be resting on a post, in which case it is pushing down on the post which is being compressed, or it may be hanging from a rod so that the rod is being stretched by the pull. In dealing with calculations involving forces we must always be quite clear whether the force is causing *compression* by pushing or *tension* by pulling. You will find this very important when you deal with the mechanics of force in your science lessons.

Although in building construction, we deal with static forces, the true definition for force involves motion. Sir Isaac Newton was a great scientist and established the truth of many basic principles. Newtons first law of motion states that, 'Every body continues in a state of rest or uniform motion in a straight line unless compelled by some external force to alter that state'.

From this we get the standard definition of force which is; 'Force is that which changes, or tends to change, the state of rest of a body or its uniform motion in a straight line'.

The unit of force is that force required to produce an acceleration of one metre per second per second (m/s^2) in a mass of one kilogram (1 kg). This is a compound unit of $1 \, kg \, m/s^2$ but it has a special name, the *newton*. Its symbol is N. (All units named after people have capital letters as their unit symbols.) When you study mechanics to greater depth you will understand that the connection between mass and force is the gravitational constant, g. In SI units this is $9.81 \, m/s^2$. That is:

$1 kgf = 9.81 \, N$

(Note the 'f' which differentiates between a mass of 1 kilogram and the force of gravity which acts on it.)

From the above connection you can see that the newton is a very small force. Consequently the multiple unit, the kilonewton (kN), is useful. This is of course, one thousand newtons.

When a force acts over an area *pressure* is created. Pressure is measured as 'force per unit area' and is therefore expressed in units such as 'newtons per square millimetre' (N/mm^2) or kilonewtons per square metre (kN/m^2), etc.

Example 1

A stack of 1500 bricks covers a ground area 2.750 m by 1.350 m. The average mass of the bricks is 3.200 kg. Calculate the total force, and the pressure, on the ground under the stack.

Force on ground = total weight of stack
$$= 1500 \times 3.2 \times 9.81 \text{ N}$$
$$= 47090 \text{ N}$$
$$= 47.09 \text{ kN}$$

Pressure $= \dfrac{\text{force}}{\text{area}}$

$$= \frac{47.09 \text{ kN}}{2.75 \times 1.35 \text{ m}^2}$$
$$= 12.69 \text{ kN/m}^2 \text{ (kilonewtons per square metre)}$$

No.	Log
2.75	0.4393
1.35	0.1303
	0.5696
47.09	1.6729
	0.5696
12.69	1.1033

It is worth noting here that the ground under the stack of bricks must be resisting this pressure with an *equal* and *opposite* reaction. This resistance is called *stress*. Since stress is equal and opposite to pressure it is therefore expressed in the same units.

Example 2

A 100 mm diameter pipe is fitted with a plug sealing one end. The force exerted on the plug by water in the pipe is 3 kN. Calculate the pressure on the plug in newtons per square millimetre (N/mm^2).

$$\text{Pressure} = \frac{\text{force}}{\text{area}}$$

$$= \frac{3000 \text{ N}}{\pi(50)^2 \text{ mm}^2}$$

$$= \frac{3000}{2500\,\pi} \text{ N/mm}^2$$

$$= \frac{6}{5\pi} \text{ N/mm}^2 \text{ (after cancelling)}$$

$$= 0.3819 \text{ N/mm}^2$$

No.	Log
5	0.6990
π	0.4972
	1.1962
6	0.7782
5π	1.1962
0.3819	1.5820

Example 3

A steel column weighing 181.5 kgf and carrying a load of 1500 kg mass rests on a concrete foundation. The base plate between the column and the foundation has an area of 645 cm^2. Find the total force on the concrete in kN and the pressure immediately beneath the plate in kN/m^2.

$$\text{Total force on concrete} = \text{weight of column} + \text{load}$$

$$= 181.5 + 1500 \text{ kgf}$$

$$= 1681.5 \times 9.81 \text{ N}$$

$$= \frac{1681.5 \times 9.81 \text{ kN}}{1000}$$

$$= 16.5 \text{ kN}$$

$$\text{Pressure beneath plate} = \frac{\text{force}}{\text{area}}$$

$$= \frac{16.5}{645} \text{ kN/cm}^2$$

This gives the pressure in kN/cm^2 but we want it in kN/m^2. It will be greater sinces a square metre is 100^2 square centimetres.

$$\text{Pressure} = \frac{16.5}{645} \times 100^2 \text{ kN/m}^2$$

$$= 255.8 \text{ kN/m}^2$$

No.	Log
16.5	1.2175
100^2	4.0000
	5.2175
645	2.8096
255.8	2.4079

Work

Work is done when a force moves through a distance. The amount of work done is measured as *force x distance.*

Since newtons are the units of force and metres the units of distance it follows that work will be measured in newton metres (Nm). To do work is to use energy and therefore energy has the same units as work. In fact all forms of energy are measured in this same unit. However, like the unit of force, the work or energy unit has a special name. It is the joule, symbol J. That is:

1 Nm = 1J

The joule is used also to measure heat and electrical energy. You may be thinking that this is more appropriate to mechanical engineering apprentices than building apprentices, but you will find that an ever increasing process of mechanisation will make it more and more advantageous for you to understand the units of force and motion. Lifting hoists, pipe and bar bending machines, concrete vibrators, and mechanised hand tools are but a few of the examples to which this applies.

Example 4

How much work is done when a load of 50 kg is lifted on a block and tackle to a scaffold platform 3.600 m high?

Work done = force x distance
            = 50 x 9.81 x 3.6 Nm
            = 1766 Nm

Note: The 50 kg mass is converted to newtons (force) by multiplying by 9.81.

But a newton metre is a joule, therefore:

Work done = 1766 J
            or 1.766 kJ (since 1000 joules = 1 kilojoule)

No.	Log
50	1.6990
9.81	0.9917
3.6	0.5563
1766	3.2470

This amount of work is a measure of the energy actually expended on the load of 50 kg. The person operating the lifting tackle would, however, have to use extra energy over and above this amount to overcome the inefficiency of the apparatus. Friction between moving surfaces uses up energy and may account for quite a large amount in some cruder forms of machinery.

The man pulling on the rope or chain of the block and tackle in Example 4 would therefore use more energy than the 1.766 kJ calculated as having been applied to the load. If we describe the energy he uses as *work put in* and the energy applied to the load as *work got out* the difference between these two quantities is the energy wasted in the apparatus.

The ratio of $\dfrac{\text{work got out}}{\text{work put in}}$ is a measure of the efficiency of any machine and, when multiplied by 100, gives the percentage efficiency, that is:

$$\text{Efficiency} = \frac{\text{work got out}}{\text{work put in}} \times 100\%$$

Example 5

A load of 100 bricks of average weight 2.6 kgf is raised through a height of 4.500 m using a hand winch. The average force applied to the winch handle by the operator is 100 N and during the operation his hand moves through a distance of 210 m. What is the efficiency of the winch?

Work got out = load x distance load moves
= (100 x 2.6 x 9.81) x 4.500 J
= 11480 J
(or 11.48 kJ)

No.	Log
100	2.0000
2.6	0.4150
9.81	0.9917
4.5	0.6532
11480	4.0599

Work put in = effort x distance effort moves

= 100 x 210 J

= 21000 J (or 21 kJ)

Note that in the calculation of *work got out* the weight of the bricks in kgf was multiplied by 9.81 to convert to newtons. This did not apply in the calculation of *work put in* since the effort was given in newtons.

Efficiency $= \dfrac{\text{work got out}}{\text{work put in}} \times 100\%$

$= \dfrac{11.48}{21} \times 100\%$

$= 54.68\%$

No.	Log
11.48	1.0600
100	2.0000
	3.0600
21	1.3222
54.68	1.7378

Equilibrium

Equilibrium is a state of balance or stability. Two tug of war teams pulling with exactly equal forces are in equilibrium; so is a seesaw with its two sides exactly balancing.

In your science lectures you will have learned that for a state of equilibrium, forces in exactly opposite directions must be equal (that is, upwards and downwards or left and right) and that the clockwise turning effects of the forces must be equal to the anticlockwise turning effects.

The turning effect of a force is measured as a *turning moment*, and is the product of the *force* and the *perpendicular distance* to its line of action from the point about which turning is being considered. This distance is called the 'arm' of the moment.

Example 6

A force of 220 N is applied to the end of a single handled wrench 300 mm long. Calculate the resulting turning moment in N m (see Figure 111).

Moment = force x arm

= 220 x 0.300 N m

= 66 N m

Note that the arm of 300 mm has been used as 0.300 m. The newton metre is a compound unit and should be written with a space between its two component units that is, N m, as should all compound units. Since it is a measure of a rotational effect the units are *not* joules.

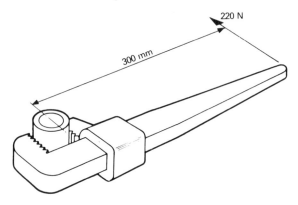

Figure 111

Example 7

The system of forces shown in Figure 112 is not in equilibrium. What horizontal and vertical forces are required to produce equilibrium?

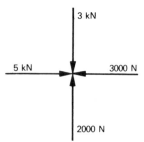

Figure 112

First put all four forces in the same units (see Figure 113).

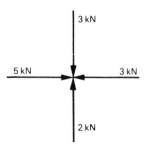

Figure 113

For equilibrium, forces to left = forces to right, and forces up = forces down.

Horizontal force required = 5 - 3 kN
                            = 2 kN to left
Vertical force required   = 3 - 2 kN
                            = 1 kN upwards

Example 8

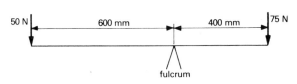

Figure 114

Is the system of forces shown in Figure 114 in equilibrium?

This may be thought of as a seesaw. For equilibrium about the fulcrum F, the clockwise turning moments must be equal to the anticlockwise turning moments

Clockwise moment   = 75 × 400
                       = 30000 N mm
Anticlockwise moment = 50 × 600 N m
                       = 30000 N mm

Since they are equal the system is in equilibrium about F.

Example 9

What force is required at x in Figure 115 to produce equilibrium?

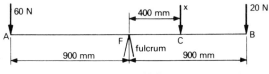

Figure 115

For equilibrium, the clockwise turning effect about the fulcrum F due to the 20 N force and the unknown force, x, must be balanced by the anticlockwise turning moment due to the 60 N force. This can be written as an equation. That is:

$(20 × 900) + (x × 400) = (60 × 900)$
$18000 + 400x \qquad\qquad = 54000$

Subtracting 18000 from each side:

$400x = 54000 - 18000$
$\quad\quad = 36000$

Dividing both sides by 400:

$$x = \frac{36000}{400}$$
$$= 90 \text{ N}$$

Exercise 18

1   The mass of material carried by a concrete foundation slab bearing a steel column is 60 tonnes. The slab itself weighs 4500 kgf. Calculate the pressure on the earth under the slab which is 1.800 m square. (1 tonne = 1000 kg; 1 kgf = 9.81 N.)

2   A press for forming sheet steel brackets applies a force of 300 kN to sheet metal over a circular area of diameter 350 mm. What is the resulting pressure in N/mm^2?

3   A concrete cube was subjected to a compressive load in a testing machine until it was crushed. At breaking point the load was 495 kN. Calculate the size of the cube if the breaking stress (pressure) was 22 N/mm^2.

4   How much work is done by the application of a force of 520 kN to push a mandril into a steel tube a distance of 75 mm? Give the answer in kJ.

5   The energy required to operate a set of pulley blocks and lift a casting weighing 120 kgf through a height of 2.400 m is 4 kJ. What is the efficiency of the pulley blocks?

6   A man loading a hoist platform lifts 72 concrete blocks, each weighing 12.5 kgf, through a height of 600 mm. How much work does he do in Nm; and in kJ?
    He then operates the hoist and the load is carried up to the fourth storey a distance of 9.600 m. How much energy is required to raise the hoist if its own weight is one tenth of the load on it and it is 65% efficient?

7   A uniform beam is pivotted at its mid point so that it is in equilibrium. A mass of 14.5 kg is hung from it at a point 1.500 m from the pivot and on its left. A mass of 9 kg is hung at a point 3.000 m to the right of the pivot. Is the beam still in equilibrium?

8   What is meant by the 'moment of a force' about a point? How is it measured?
    A force of 90 N is applied to the rim of a valve operating wheel of radius

450 mm. What is the moment of the force about the centre of the wheel?

How much energy is used when this force turns the wheel through three complete revolutions?

9   A steel rod 1.200 m long is used as a lever to lift a concrete slab weighing 400 kgf. The lower end of the rod is used as the fulcrum and the slab is supported at a point 75 mm from that end. What force is needed at the other end of the rod to lift up the slab? Give the answer in N.

10   Find what forces are needed and where they should be applied, to produce equilibrium in each of the following situations.

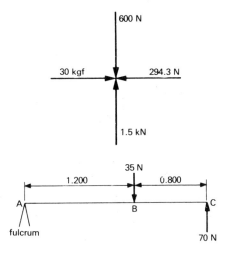

Figure 116

12 Cost calculations

Cost calculations are a very important part of the builder's work and the success of a business can depend greatly on accurate estimating. Those who try to price a job by guesswork will often find that their estimate is too high for acceptance or so low that a reasonable profit cannot be made. Experienced estimators know that 'guesstimates' should only be used as rough guides.

The work covered in this chapter deals with simple jobs or parts of jobs. In some cases rates include profits while in others a percentage is added for profit. The method may be divided into three sections:

1   Measurement of the work.
2   Pricing of the labour and materials.
3   Additions for profits and overheads.

The examples given include simple calculations for the costs of such things as areas of glass for glazing, decorations of wall and ceiling areas and volumes for excavation. Some slightly more complex calculations for brick walling, including excavation and foundations, and for roof coverings are also shown.

Some of the examples include the build-up of a unit cost and evaluations by the 'method of proportional parts' is used in several cases.

Unit costs or rates often include both labour and materials but sometimes labour costs are treated separately as shown in other examples.

Example 1

Find the cost of a plywood panel measuring 2.700 m by 1.500 m at £4.54 per square metre.

Area = 2.700 x 1.500 m²	2.7	4.54
= 4.05 m²	1.5	4.05
Cost = 4.05 x £4.54	2.7	18.16
= £18.39 (nearest 1p)	1.35	0.2270
	4.05	18.387

Example 2

Each of the six panes of glass in a window measure 600 mm by 400 mm. Find the total area of glass in square metres and its cost at £10.40 per square metre.

Total area = 6(0.600 x 0.450) m²
           = 1.62 m² (converting the dimensions to metres)

Cost = 1.62 x £10.40
     = £16.848

     = £16.85 (nearest 1p)

```
  1.62
 10.40
 16.2
  .648
16.848
```

Example 3

A pit excavated for a rainwater soakaway is 1.800 m square on plan and 1.500 m deep. How many cubic metres of earth are removed and what would the excavation cost at £15.64 per cubic metre?

Volume removed = 1.8 x 1.8 x 1.5 m³
               = 4.86 m³
Cost = 4.86 x £15.64
     = £76.0104
     = £76.01 (nearest 1p)

```
 1.8
 1.8
 1.8
 1.44
 3.24
 1.5
 3.24
 1.62
 4.86
```

```
15.64
 4.86
62.56
12.512
 .9384
76.0104
```

Example 4

A kitchen working top 3.150 m long and 650 mm wide has to be surfaced with laminated plastic at a cost of £17.50 per square metre. Find the cost of the work.

Area = 3.15 x 0.650 m²
     = 2.0475 m²
Cost = 2.0475 x £17.50
     = £35.83 (nearest 1p)

```
 3.15
 0.65
 1.890
 0.1575
 2.0475
```

```
 2.0475
17.5
20.475
14.3325
 1.02375
35.83125
```

Example 5

Ready-mixed concrete supplied at £29.20 per cubic metre is used for the base slabs of twelve garages, each slab measuring 5.000 m long, 2.500 m wide and 150 mm thick. Calculate the cost of the concrete.

Volume of concrete per slab $= 5.000 \times 2.500 \times 0.150$ m^3
$$= 1.875 \text{ m}^3$$
Total volume $= 1.875 \times 12$ m^3
$$= 22.5 \text{ m}^3$$
Cost $= 22.5 \times £29.20$
$$= £657$$

Method of proportional parts

The calculation of costs may often be simplified by subdividing the item into proportional sections and, working from a unit cost, arriving at the total cost by the addition of those sections. Any allowances such as discounts, percentage increases or reductions should be calculated at the final stage.

Example 6

Find the cost of 285 kg of sheet lead at £968 per tonne.
Note: 1 tonne = 1000 kg.   £

Unit cost	1 tonne (1000 kg) @ £968 = 968.00
($\frac{1}{4}$ of previous line)	250 kg @ £968 = 242.00
($\frac{1}{10}$ of previous line)	25 kg @ £968 = 24.20
($\frac{1}{100}$ of unit cost)	10 kg @ £968 = 9.68
(Total)	285 kg @ £968 = £275.88

Notice that the unit cost written in the first line is underlined. This is a reminder *not to include it* in the final addition.

Example 7

Find the cost of 25,250 facing bricks at £145 per thousand to be purchased at a trade discount of 10% plus $2\frac{1}{2}$% of the discounted price for prompt payment.

Unit cost	1000 @ £145 = £145
	25000 @ £145 = £3625.00
($\frac{1}{100}$ of previous line)	250 @ £145 = £36.25
(Total)	25250 @ £145 = £3661.25
Less 10% trade discount ($\frac{1}{10}$)	£366.125
	£3295.125
Less $2\frac{1}{2}$% ($\frac{1}{40}$) (nearest 1p)	£82.38
Final cost	£3212.745

To nearest 1p £3212.75

Notice the alternative methods of writing the £ sign. In Example 6 it is written once above the column. In Example 7 it is used before every amount written down. The first way seems the most appropriate to the 'practice' method.

Example 8

Find the total cost of preparing and painting three coats of oil colour on 225 m^2 of plastered wall surface at the following rates:

Preparing surface and applying first coat	£1.50p per m^2
Each additional coat	68p per m^2

First build up the overall rate per square metre to find the unit cost.

	£
Prepare and apply one coat	1.50
Two further coats @ 68 p	1.36
Unit cost per m^2	2.86
200 m^2 @ £2.86	572.00
25 m^2 @ £2.86	71.50 ($\frac{1}{8}$ of previous line)
225 m^2 @ £2.86	£643.50

Example 9

Calculate the cost of excavating the rainwater soakaway in Figure 117 filling it to within 300 mm of the top with broken brick hardcore and covering with a 300 mm layer of earth. Use the price rates given below and add 5% for overheads (the rates include profit).

Excavation (by hand)	£31.13 per m^3
Hardcore fill	£11.11 per m^3
Returning earth 300 mm thick	£1.60 per m^2

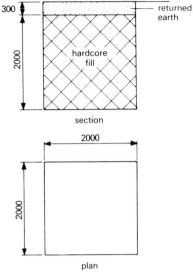

Figure 117

Measurement:

Volume for excavation = $2 \times 2 \times 2.3$ m^3

                              = 9.2 m^3

Volume of hardcore   = $2 \times 2 \times 2$ m^3

                              = 8 m^3

Area of cover   = 2×2 m^2

                              = 4 m^2

Pricing: £

Excavation 9.2 m^3 @ £31.13 =   286.40

Hardcore 8 m^3 @ £11.11   =   88.88

Cover 4 m^2 @ £1.60   =   6.40

Total for labour and materials   381.68

Additions:

Add 5% for overheads ($\frac{1}{20}$)   19.08

Total estimated price   400.76

Note that the £ sign is placed at the top of the column to save repeating it on each line.

Example 10

Estimate the cost of the concrete footings and brickwork for a 7.500 m length of the wall shown in Figure 118. Allow 30% for profit and overheads combined. Use the following rates.

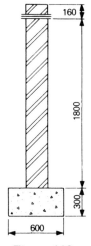

Figure 118

Concrete in footings	£34.84 per m³
225 mm brickwork	£18.87 per m²
Brick on edge coping (including tile creasing	
and weather fillets)	£4.27 per m run

Measurement:
Volume of concrete = 7.5 x 0.6 x 0.3 m³
                   = 1.35 m³
Area of brickwork   = 7.5 x 1.8 m²
                   = 13.5 m²
Length of coping   = 7.5 m

Pricing: £
Concrete   1.35 m³ @ £34.84 =   47.03
Brickwork   13.5 m² @ £18.87 = 254.75
Coping   7.5 m @ £4.27   =   32.03
Total for labour and materials = 333.81

Additions:
30% for profit and overheads
(30% = $\frac{3}{10}$) $\frac{3}{10}$ x 333.81 = 100.14
Total estimated cost   433.95

Example 11

A flat roof measuring 6.750 m by 3.300 m is to be boarded and covered with three-ply roofing felt. The fascia boards are 150 mm deep and of the same thickness as the roofing boards. Estimate the cost of boarding and felting if the boarding costs £7.14 per m² and the felt (three-ply) £7.15 per m² to supply and fix. Allow 20% for overheads and profit and 10% waste on the felt for laps and cutting.

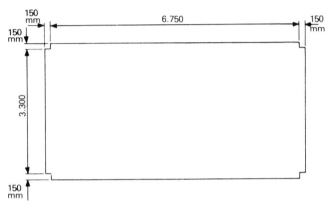

Figure 119

Measurement:

If the fascia boards were turned up we would have the shape shown in Figure 119. The total area of boarding and felting is therefore the area of the overall rectangle minus the four corners. The area of felt required is 10% greater.

Area of each corner = 150 x 150 mm²
                      or 0.15 x 0.15 m²
Total deduction   = 4(0.15 x 0.15) m²
                   = 0.09 m²
Total length   = 6.750 + 2(0.150) m
                   = 7.050 m
Total width   = 3.300 + 2(0.150) m
                   = 3.600 m
Area for boarding = (7.050 x 3.600) - 0.09 m²
                   = 25.38 - 0.09 m²
                   = 25.29 m²

Area for felting

add 10% to above ($\frac{1}{10}$) $\frac{2.53 \text{ m}^2}{27.82 \text{ m}^2}$ (correct to two places)

Pricing:

Boarding   25.29 m² @ £7.14
            25.29 x 7   = 177.03
            25.29 x 0.1   =   2.53
            25.29 x 0.04 =   1.01
Total   180.57   £180.57

Felting   27.82 m² @ £7.15
            27.82 x 7   =   194.74
            27.82 x 0.1   =   2.78
            27.82 x 0.05 =   1.39
Total   198.91   £198.91
Total for labour and materials   379.48

Additions:
20% for overheads and profit ($\frac{1}{5}$)   75.90
Total estimated cost   £455.38

Example 12

A room is to be decorated using wallpaper at £3.42 per piece and ceiling paper at £1.25 per piece (a roll of paper is called a piece in the decorating trade). The room is 4.200 m long, 3.150 m wide and 2.400 m high from floor to ceiling. The frieze is 325 mm deep and the skirting 225 mm high. Allowing $12\frac{1}{2}$% waste estimate the cost of materials only, assuming that one piece covers 5.50 m².

Measurement:
Perimeter of room   = 2(4.200 + 3.150) m
                          = 2 x 7.350 m
                          = 14.700 m
Height of walls   = 2.400 – (0.225 + 0.325) m
                          (total less skirting and frieze)
                          = 2.400 – 0.55 m
                          = 1.85 m
Area of walls   = perimeter x height
                          = 14.700 x 1.8500 m²
                          = 27.195 m²
Add $12\frac{1}{2}$% waste ($\frac{1}{8}$)   3.400 m²
Total   30.595 m² say 30.60
Number of pieces   = 30.60 ÷ 5.5
                          = 5.563, say 6 pieces.

The above division is best carried out by factors. First make the divisor a whole number, that is:

$$\frac{30.6}{5.5} = \frac{306}{5.5}$$

$$5 \underline{)\ 306\ \ \ \ \ }$$
$$11\ \underline{)\ 61.2\ \ \ }$$
$$5.563$$

Now divide by 5 and then by 11.

Area of ceiling   = 4.200 x 3.150 m²
                             = 13.23 m²
Area of frieze   = 14.700 x 0.325 m² (perimeter x frieze depth)
                             =   4.78 m²

Total area for ceiling paper   = 13.23 + 4.78 m^2

                                      = 18.01 m^2

Add 12$\frac{1}{2}$% waste ($\frac{1}{8}$)   2.25 m^2

Total   20.26 m^2

Number of pieces   = 20.26 ÷ 5.5

                                        = 3.684, say 4 pieces

Pricing: £

6 pieces @ £3.42 = 20.52

4 pieces @ £1.25 = 5.00

                                 25.52

Example 13

The installation of two radiators to be connected to an existing heating system involves the following in labour and materials: two radiators at £15.000 each, 9.600 m of 22 mm copper tube at £1.27 per metre, six 22 mm bent couplings at 22p each, three 22 mm tees at 63p each, ten hours of labour by a craftsman at 93p per hour and his apprentice at 56p per hour.

Estimate the total cost of the work allowing 25% for overheads and profit.

In this problem no calculations involving measurement are required.

Pricing:

Materials			£	
Radiators	2 @ £15.000	=	30.00	
Pipe	9.60 @ £1.27	=	12.19	
Fittings	6 @ 22p	=	1.32	
	3 @ 63p	=	1.89	£
Total materials			45.40	45.40

Labour:		£	
Craftsman 10 hour @ 93p	=	9.30	
Apprentice 10 hour @ 56p	=	5.60	
Total labour		14.90	14.90
Total labour and materials			60.30

Additions:	
Overheads and profit 25% = $\frac{1}{4}$	15.07$\frac{1}{2}$
Total estimated cost	75.37$\frac{1}{2}$

Exercise 19

1   Calculate the cost of preparing and applying two coats of emulsion paint on 111 m^2 of plastered walls at the following rate: preparing and applying one coat emulsion 70p per m^2; each subsequent coat 45p per m^2.

2   The walls of a room, which is 3.750 m long and 3.250 m wide, have a height of 1.875 m between the top of the skirting and the lower edge of the lower edge of the picture rail. How many m² of wall surface are there for papering, if 20% of this wall space is taken up by door and window openings? What will be the cost at £1.50 per square metre?

3   An apprentice receives 138p per hour for an 8½ hour day, working 5½ days per week.
   (a)   What is his gross weekly wage (excluding deductions)?
   (b)   What will his total weekly *increase* be next year when he receives a 20% rise?

4   What is the total weekly gross wage packet for a gang of four bricklayers working a 48-hour week at 173p per hour?
   Note: Gross means before deductions are made.

5   Find the cost of 165 m² of tongued and grooved boarding supplied and fixed at £97.20 per 10 m².

6   Calculate the cost of 895 m of p.v.c. guttering at £2.91 per metre.

7   What is the cost of hacking out sixteen panes of broken sheet glass, each pane measuring 1200 mm by 450 mm, at £3.42 per square metre?

8   Calculate the cost of reglazing the frames of question 7 at £11.24 per square metre?

9   What would be the cost of supplying and fixing six birch plywood panels, each measuring 2.250 m x 1.200 m at £12.15 per square metre plus 8% overheads.

10   A lean-to roof measures 7.500 m in length and 4.800 m along the line of greatest slope. Calculate the cost of slating it at £30.68 per m².

11   A damaged door is to be replaced including new hinges and lock. If the door costs £35.80, the hinges and lock £5.25 and the job takes a joiner 3 hours at 173p per hour, calculate the cost of the work allowing 25% for overheads and profit.

12   A dwarf boundary wall 750 mm high and 112.5 mm thick is to be built on an existing concrete slab. The wall is to be pointed on both sides and finished along the top with a cement and sand fillet. Calculate a price for the work using the following rates, which include profit, and allowing 8% for overheads.

112.5 mm brickwork	= £14.50 per m²
Raking out and pointing joints	= £1.84 per m²
Cement and sand fillet	= £1.75 per m run

13   Figure 120 shows the plan and elevation (in outline only) of a simple roof. Using the rates given below, which include the cost of materials and labour

only, estimate the cost of boarding, felting, battening and tiling the roof. Add 25% for profit and 5% for overheads.

Boarding   =   £7.10 per m²
Roofing felt   =   £2.72 per m²
Tiling slopes (including battens)   =   £16.88 per m²
Ridge tiling   =   £4.16 per m run

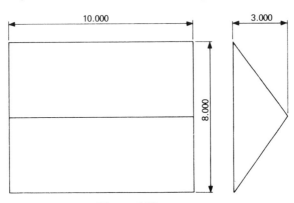

Figure 120

14   Using the same rates as were used in Example 9 on page 177, find the cost of a soakaway constructed in a similar manner to that in Figure 117 but circular on plan with a diameter of 2.000 m and a depth of 2.500 m. Allow 8% for overheads.

15   A room is 4.500 m long, 3.600 m wide and 2.700 m high from floor to ceiling. It has two doors each 2100 x 800 mm measured over the frames and one window 2400 x 1200 mm overall. The skirting is 150 mm high. Estimate the cost of decorating the room as follows:

Ceiling to be painted 2 coats emulsion at £1.15 per m²
Walls to be prepared and painted 2 coats at £1.25 per m²
Doors to be prepared and painted at £2.45 per m²
Skirting to be prepared and painted at 64p per m run
Floor to be prepared and stained and varnished at £3.38 per m²

   Allow 25% for overheads and profit.

16   Calculate the cost of the timber at £187.10 per m³ for a job requiring the following:

Twenty four 3.600 m lengths of 50 mm x 180 mm
14.4 m run of 50 mm x 75 mm
3.6 m run of 75 mm x 200 mm

17   Using the following rates calculate the cost of excavating for and laying a

20.000 m run of 100 mm diameter drain pipe on and including a 150 mm concrete bed benched up to the top of the pipe. Allow 33⅓% for profit and overheads. Assume the excavation to average 1.500 m in depth.

Excavation of trench average 1.500 m deep,	
backfilling and disposing of surplus earth	£10.42 per m run
Concrete bed and benching	£2.38 per m run
Drain pipes laid and jointed	£2.49 per m run

18 A new window opening 1200 x 900 mm is to be made in a 225 mm wall. The work involves cutting out the brickwork at £2.75 per m², constructing a tile sill externally at £2.95 per m run and a quarry tile sill internally at £3.25 per m run and providing and fixing a reinforced concrete lintel 1500 mm long at £7.25 per m run. Additional items are making good the reveals at £2.40 per m run and supplying and fixing the frame at a cost of £92.50.

    Allow 30% for overheads and profit and calculate the cost of the work.

Revision exercise 3

1 The formula $A = \pi(R - r)(R + r)$ gives the area of a flat ring (or washer) with an external radius of R, and an internal radius of r, units. Find the area in mm² of such a ring having an outside diameter of 500 mm and an inside diameter of 200 mm.

2 In the formula $S = 2\pi r(h + r)$, S is the total surface area of a closed or solid cylinder of radius r, and height h, units. Find in m², correct to two places of decimals, the surface area of a cylinder of 525 mm diameter and 700 mm height.

3 Find the value of P from the formula
$$P = \sqrt{\frac{R - S^2}{R + S^2}},$$
when $R = 65$ and $S = 4$.

4 The capacity, C, of a tank in litres is given by the formula $C = 1000lbd$, where l, b and d are the length, breadth and depth in m. Find the capacity of a tank of length 1.350 m, breadth 1.900 m and depth 825 mm, giving your answer correct to four significant figures.

5 Figures 121, 122 and 123 show three different structural sections, a channel, a tee and an angle. Denote the thickness by t in every case. Using the symbols given obtain a formula for the area of each.

6 Obtain an expression (formula) for the weight, W, in grams of a rectangular sheet of lead weighing 39 kg per m² in terms of its length, a mm, and its breadth, b mm.

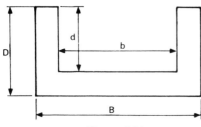

Figure 121

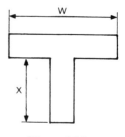

Figure 122

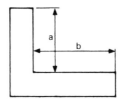

Figure 123

7   Taking the density of concrete as 2240 kg/m³, find an expression (formula) for the weight, W, in kg of a concrete slab x m long, y m wide and z mm thick.

8   (a)   A concretor making precast lintels in paid P pence per hour plus a bonus of 50p for every ten lintels cast. Obtain a formula for his total pay, W, in pounds, in a 44 hour week during which he produces N lintels.

(b)   Use the formula to calculate the concretor's pay for a week in which his rate is 180p per hour and during which he makes 50 lintels.

9   Calculate the value of P from the formula,

$$P = \left(1 - \frac{b}{s}\right) 100\%,$$

when $b = 2147$ and $s = 2666$.

10   (a)   Construct a formula for the surface area of a solid half-cylinder formed by cutting a cylinder in two through its vertical centre line. Use the usual symbols, h for height and r for radius.

(b)   What is the surface area of such a solid having a radius of 75 mm and a height of 175 mm? Give the answer in cm².

11   The formula, $V = \frac{1}{3}\pi r^2 h$ gives the volume of a cone of height h and with a base of radius r. This can be rearranged to give r in terms of V and h, that is:

$$r = \sqrt{\frac{3V}{\pi h}}$$

Find the radius of the base of a cone of height 80 mm whose volume is 72 cm^3.

12   A brick has a weight of w_1 kg when dry and w_2 kg when saturated with water. Express the percentage increase in weight, P, in terms of w_1 and w_2.
 What is the percentage increase in weight of a brick weighing 2.5 kg dry and 3.125 kg when saturated?

13   Plot on graph paper the four points (–3, –2), (1, –2), (–1, 4) and (3, 4). What is the area of the largest figure formed by joining them?

14   In the following table are given the building craftsman's rates of pay for the years 1950 to 1960. Plot a graph showing the trend of these rates over the ten year period. (A look back at the old £ - s -d system!)

Year	1950	1951	1952	1953	1954	1955
Rate	3s 0d	3s 3d	3s 6d	3s 8d	3s 9d	3s 11d
Year	1956	1957	1958	1959	1960	
Rate	4s 2½d	4s 6d	4s 8½d	4s 10½d	5s 1d	

15   In the following table are values of pressure measured in kN/m^2 at various depths in a test tank of water.

Depth (m)	0	0.500	1.000	2.000	3.000	4.500	6.000
Pressure (kN/m^2)	100	104.91	109.81	119.62	129.43	144.15	158.86

Plot the points very carefully and draw the average straight line through them. From you graph read off:
(a)   The pressure at a depth of 4.000 m.
(b)   The depth at which the pressure is 135 kN/m^2.
(This unit of pressure is 'kilonewtons per square metre'.)

16   In an experiment of concrete strength, test cubes are tested at the ages shown in the table and their crushing strengths in N/mm^2 obtained. Plot these values on a graph, using the vertical axis for strengths and join them with a smooth curve. From your graph obtain:
(a)   The likely strength of this type of concrete at an age of 15 days.
(b)   The age at which the strength is likely to reach 21 N/mm^2.

Age (days)	3	7	14	21	28
Strength (N/mm²)	9.45	15.75	22.40	25.90	28.00

17   Draw a straight line graph showing the relationship between the length and the weight of 18 mm lead pipe, which weighs 4.40 kg per metre. Use the graph to find the weight of a 2.750 m length of the pipe. What would be the length of 7.5 kg of the pipe?

18   Taking 2.205 lb as being the equivalent of 1 kilogram (1000 grams) draw a straight line graph for converting grams and ounces up to a value of 1 lb. From your graph convert:
(a)   10 oz to grams.
(b)   350 g to ounces.

19   If the hourly rate of payment for a craftsman is 180p, draw a graph from which you can read off his weekly earnings for any number of hours up to 44, and then for overtime at time and a half up to a total of 60 hours. From you graph obtain:
(a)   The wage of a man who works a 40 hour week
(b)   The wage of a man who works 54 hours in a week (overtime after 44 hours).
(c)   The number of hours worked by a man whose wage is £90.00.

20   A load of mass 750 kg is supported on a square platform resting on the ground. The platform is 175 mm square. What is the resulting pressure on the ground in kN/m²?

21   In a test for evaluating the compressive strength of a cement and sand mortar a prepared cube of the material, of edge length 7.071 cm, was subjected to a compressive load to failure. The ultimate load was 150 kN. What was the pressure on the face of the cube at the moment of failure?

22   The force required to lift a loaded goods lift through a height of 2.400 m is 2 kN. What work is done in this operation?

23   The energy supplied to a lifting machine when it lifts a load of mass 240 kg through a height of 1.200 m is 3.85 kJ. What is its efficiency?

24   A man applies a force of 250 N to a winch handle of radius 500 mm and turns it through 150 revolutions in raising a load of weight 250 kgf through a height of 15 m. Calculate:
(a)   The work done by the man.
(b)   The work done on the load.
(c)   The efficiency of the winch.

25   A floor to be tiled in 150 mm square quarry tiles is shown in Figure 124. Find:

(a)   The number of tiles required, allowing 5% waste.

(b)   Their cost at £19.20 per hundred.

26   A building contains a total area of 192 m² of brickwork 225 mm thick. Allowing 7½% for cutting and breakages, how many bricks are required and what will they cost at £136.50 per 1000? Allow a brick size to be 225 mm x 112.5 mm x 75 mm.

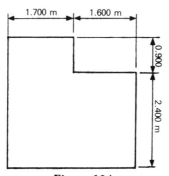

Figure 124

27   The cost of matt emulsion for wall decoration is £9.50 per 5 litres and of floor paint £15.25 per 5 litres. Assume the covering capacity of the wall emulsion to be 24 m² per litre and that of the floor paint 18 m² per litre. Calculate the cost of these materials for a hall 36 m long, 26 m wide and 6 m high wall to ceiling, allowing a deduction of 25% of the wall area for door, window and fireplace openings.

28   A wall surface 2.550 m long and 1.800 m high is to be tiled with 100 mm square tiles. The cost of the tiles is £5.85 per 100. The work consists of preparing the surface at 90p per m² and fixing the tiles at £6.10 per m². Estimate the total cost of the work, adding 20% for profit.

29   A floor is to be laid in T & G boards 25 mm thick. The floor is 5.400 m long and 3.750 m wide. The cost of the work is £9.65 per m² for laying the floor and £4 per m² for finishing and staining. Find:

(a)   The total cost of the work.

(b)   The weight of wood used if its density is 850 kg/m³.

30   In a public building a total length of 450 m of corridors, all 1.250 m wide, are to be cleaned off at a cost of £1.50 per m² and finished with 25 mm thick granolithic paving costing £3.95 per m². Certain parts of this totalling 5% of the whole are to be finished with trowelled in carborundum at an extra cost of £1 per m². Find the total cost of the work.

31   A run of 150 mm diameter drain pipes is to be laid across the diagonal of a rectangular plot measuring 60 m x 15 m. The pipes are to be on a concrete

bed and benched up to the top. The average depth of the excavation for the work is to be 1.200 m. Using the rates below, find the cost of the work.

Excavate average 1.200 m deep, backfilling the
trench and removing surplus earth from the site   £9.74 per m run
Concrete bed and benching up pipes   £4.45 per m run
150 mm diameter pipes laid and jointed   £4.25 per m run

32   A rectangular-on-plan building has a roof pitched at 45° with gable ends. The plan dimensions are 15 m x 8 m. Using the given rates estimate the cost of tiling the roof:

25 mm boarding to slopes   £7.01 per m²
Roofing felt   £2.81 per m²
25 mm x 20 mm tile battens   £1.10 per m²
Sussex handmade tiles   £31.50 per m²
Ridge tiling   £9.50 per m run
Extra for eaves course   £4.72 per m run

33   The cost of re-finishing a floor to a room 4.500 m long and 3.600 m wide is based on the following rates.

25 mm wood block flooring   £34.20 per m²
Skirting, including mitres   £3.85 per m run
Extra for polishing floor   £5.60 per m²

Find the cost of the work, making an addition of 10%.

Reciprocals of numbers from 1 to 10

	0	1	2	3	4	5	6	7	8	9	Mean Differences 1 2 3	4 5 6	7 8 9
1·0	1·000	9901	9804	9709	9615	9524	9434	9346	9259	9174			
1·1	·9091	9009	8929	8850	8772	8696	8621	8547	8475	8403			
1·2	·8333	8264	8197	8130	8065	8000	7937	7874	7813	7752			
1·3	·7692	7634	7576	7519	7463	7407	7353	7299	7246	7194			
1·4	·7143	7092	7042	6993	6944	6897	6849	6803	6757	6711	5 10 14	19 24 29	33 38 43
1·5	·6667	6623	6579	6536	6494	6452	6410	6369	6329	6289	4 8 13	17 21 25	29 33 38
1·6	·6250	6211	6173	6135	6098	6061	6024	5988	5952	5917	4 7 11	15 18 22	26 29 33
1·7	·5882	5848	5814	5780	5747	5714	5682	5650	5618	5587	3 6 10	13 16 20	23 26 29
1·8	·5556	5525	5495	5464	5435	5405	5376	5348	5319	5291	3 6 9	12 15 17	20 23 26
1·9	·5263	5236	5208	5181	5155	5128	5102	5076	5051	5025	3 5 8	11 13 16	18 21 24
2·0	·5000	4975	4950	4926	4902	4878	4854	4831	4808	4785	2 5 7	10 12 14	17 19 21
2·1	·4762	4739	4717	4695	4673	4651	4630	4608	4587	4566	2 4 7	9 11 13	15 17 20
2·2	·4545	4525	4505	4484	4464	4444	4425	4405	4386	4367	2 4 6	8 10 12	14 16 18
2·3	·4348	4329	4310	4292	4274	4255	4237	4219	4202	4184	2 4 5	7 9 11	13 14 16
2·4	·4167	4149	4132	4115	4098	4082	4065	4049	4032	4016	2 3 5	7 8 10	12 13 15
2·5	·4000	3984	3968	3953	3937	3922	3906	3891	3876	3861	2 3 5	6 8 9	11 12 14
2·6	·3846	3831	3817	3802	3788	3774	3759	3745	3731	3717	1 3 4	6 7 8	10 11 13
2·7	·3704	3690	3676	3663	3650	3636	3623	3610	3597	3584	1 3 4	5 7 8	9 11 12
2·8	·3571	3559	3546	3534	3521	3509	3497	3484	3472	3460	1 2 4	5 6 7	9 10 11
2·9	·3448	3436	3425	3413	3401	3390	3378	3367	3356	3344	1 2 3	5 6 7	8 9 10
3·0	·3333	3322	3311	3300	3289	3279	3268	3257	3247	3236	1 2 3	4 5 6	7 9 10
3·1	·3226	3215	3205	3195	3185	3175	3165	3155	3145	3135	1 2 3	4 5 6	7 8 9
3·2	·3125	3115	3106	3096	3086	3077	3067	3058	3049	3040	1 2 3	4 5 6	7 8 9
3·3	·3030	3021	3012	3003	2994	2985	2976	2967	2959	2950	1 2 3	4 4 5	6 7 8
3·4	·2941	2933	2924	2915	2907	2899	2890	2882	2874	2865	1 2 3	3 4 5	6 7 8
3·5	·2857	2849	2841	2833	2825	2817	2809	2801	2793	2786	1 2 2	3 4 5	6 6 7
3·6	·2778	2770	2762	2755	2747	2740	2732	2725	2717	2710	1 2 2	3 4 5	5 6 7
3·7	·2703	2695	2688	2681	2674	2667	2660	2653	2646	2639	1 1 2	3 4 4	5 6 6
3·8	·2632	2625	2618	2611	2604	2597	2591	2584	2577	2571	1 1 2	3 3 4	5 5 6
3·9	·2564	2558	2551	2545	2538	2532	2525	2519	2513	2506	1 1 2	3 3 4	4 5 6
4·0	·2500	2494	2488	2481	2475	2469	2463	2457	2451	2445	1 1 2	2 3 4	4 5 5
4·1	·2439	2433	2427	2421	2415	2410	2404	2398	2392	2387	1 1 2	2 3 3	4 5 5
4·2	·2381	2375	2370	2364	2358	2353	2347	2342	2336	2331	1 1 2	2 3 3	4 4 5
4·3	·2326	2320	2315	2309	2304	2299	2294	2288	2283	2278	1 1 2	2 3 3	4 4 5
4·4	·2273	2268	2262	2257	2252	2247	2242	2237	2232	2227	1 1 2	2 3 3	4 4 5
4·5	·2222	2217	2212	2208	2203	2198	2193	2188	2183	2179	0 1 1	2 2 3	3 4 4
4·6	·2174	2169	2165	2160	2155	2151	2146	2141	2137	2132	0 1 1	2 2 3	3 4 4
4·7	·2128	2123	2119	2114	2110	2105	2101	2096	2092	2088	0 1 1	2 2 3	3 4 4
4·8	·2083	2079	2075	2070	2066	2062	2058	2053	2049	2045	0 1 1	2 2 3	3 3 4
4·9	·2041	2037	2033	2028	2024	2020	2016	2012	2008	2004	0 1 1	2 2 2	3 3 4
5·0	·2000	1996	1992	1988	1984	1980	1976	1972	1969	1965	0 1 1	2 2 2	3 3 4
5·1	·1961	1957	1953	1949	1946	1942	1938	1934	1931	1927	0 1 1	2 2 2	3 3 3
5·2	·1923	1919	1916	1912	1908	1905	1901	1898	1894	1890	0 1 1	1 2 2	3 3 3
5·3	·1887	1883	1880	1876	1873	1869	1866	1862	1859	1855	0 1 1	1 2 2	2 3 3
5·4	·1852	1848	1845	1842	1838	1835	1832	1828	1825	1821	0 1 1	1 2 2	2 3 3

Reciprocals of numbers from 1 to 10

	0	1	2	3	4	5	6	7	8	9	Mean Differences								
											1	2	3	4	5	6	7	8	9
5·5	·1818	1815	1812	1808	1805	1802	1799	1795	1792	1789	0	1	1	1	2	2	2	3	3
5·6	·1786	1783	1779	1776	1773	1770	1767	1764	1761	1757	0	1	1	1	2	2	2	3	3
5·7	·1754	1751	1748	1745	1742	1739	1736	1733	1730	1727	0	1	1	1	1	2	2	2	3
5·8	·1724	1721	1718	1715	1712	1709	1706	1704	1701	1698	0	1	1	1	1	2	2	2	3
5·9	·1695	1692	1689	1686	1684	1681	1678	1675	1672	1669	0	1	1	1	1	2	2	2	3
6·0	·1667	1664	1661	1658	1656	1653	1650	1647	1645	1642	0	1	1	1	1	2	2	2	3
6·1	·1639	1637	1634	1631	1629	1626	1623	1621	1618	1616	0	1	1	1	1	2	2	2	2
6·2	·1613	1610	1608	1605	1603	1600	1597	1595	1592	1590	0	1	1	1	1	2	2	2	2
6·3	·1587	1585	1582	1580	1577	1575	1572	1570	1567	1565	0	0	1	1	1	1	2	2	2
6·4	·1562	1560	1558	1555	1553	1550	1548	1546	1543	1541	0	0	1	1	1	1	2	2	2
6·5	·1538	1536	1534	1531	1529	1527	1524	1522	1520	1517	0	0	1	1	1	1	2	2	2
6·6	·1515	1513	1511	1508	1506	1504	1502	1499	1497	1495	0	0	1	1	1	1	2	2	2
6·7	·1493	1490	1488	1486	1484	1481	1479	1477	1475	1473	0	0	1	1	1	1	2	2	2
6·8	·1471	1468	1466	1464	1462	1460	1458	1456	1453	1451	0	0	1	1	1	1	2	2	2
6·9	·1449	1447	1445	1443	1441	1439	1437	1435	1433	1431	0	0	1	1	1	1	2	2	2
7·0	·1429	1427	1425	1422	1420	1418	1416	1414	1412	1410	0	0	1	1	1	1	1	2	2
7·1	·1408	1406	1404	1403	1401	1399	1397	1395	1393	1391	0	0	1	1	1	1	1	2	2
7·2	·1389	1387	1385	1383	1381	1379	1377	1376	1374	1372	0	0	1	1	1	1	1	2	2
7·3	·1370	1368	1366	1364	1362	1361	1359	1357	1355	1353	0	0	1	1	1	1	1	2	2
7·4	·1351	1350	1348	1346	1344	1342	1340	1339	1337	1335	0	0	1	1	1	1	1	1	2
7·5	·1333	1332	1330	1328	1326	1325	1323	1321	1319	1318	0	0	1	1	1	1	1	1	2
7·6	·1316	1314	1312	1311	1309	1307	1305	1304	1302	1300	0	0	1	1	1	1	1	1	2
7·7	·1299	1297	1295	1294	1292	1290	1289	1287	1285	1284	0	0	0	1	1	1	1	1	1
7·8	·1282	1280	1279	1277	1276	1274	1272	1271	1269	1267	0	0	0	1	1	1	1	1	1
7·9	·1266	1264	1263	1261	1259	1258	1256	1255	1253	1252	0	0	0	1	1	1	1	1	1
8·0	·1250	1248	1247	1245	1244	1242	1241	1239	1238	1236	0	0	0	1	1	1	1	1	1
8·1	·1235	1233	1232	1230	1229	1227	1225	1224	1222	1221	0	0	0	1	1	1	1	1	1
8·2	·1220	1218	1217	1215	1214	1212	1211	1209	1208	1206	0	0	0	1	1	1	1	1	1
8·3	·1205	1203	1202	1200	1199	1198	1196	1195	1193	1192	0	0	0	1	1	1	1	1	1
8·4	·1190	1189	1188	1186	1185	1183	1182	1181	1179	1178	0	0	0	1	1	1	1	1	1
8·5	·1176	1175	1174	1172	1171	1170	1168	1167	1166	1164	0	0	0	1	1	1	1	1	1
8·6	·1163	1161	1160	1159	1157	1156	1155	1153	1152	1151	0	0	0	1	1	1	1	1	1
8·7	·1149	1148	1147	1145	1144	1143	1142	1140	1139	1138	0	0	0	1	1	1	1	1	1
8·8	·1136	1135	1134	1133	1131	1130	1129	1127	1126	1125	0	0	0	1	1	1	1	1	1
8·9	·1124	1122	1121	1120	1119	1117	1116	1115	1114	1112	0	0	0	1	1	1	1	1	1
9·0	·1111	1110	1109	1107	1106	1105	1104	1103	1101	1100	0	0	0	1	1	1	1	1	1
9·1	·1099	1098	1096	1095	1094	1093	1092	1090	1089	1088	0	0	0	0	1	1	1	1	1
9·2	·1087	1086	1085	1083	1082	1081	1080	1079	1078	1076	0	0	0	0	1	1	1	1	1
9·3	·1075	1074	1073	1072	1071	1070	1068	1067	1066	1065	0	0	0	0	1	1	1	1	1
9·4	·1064	1063	1062	1060	1059	1058	1057	1056	1055	1054	0	0	0	0	1	1	1	1	1
9·5	·1053	1052	1050	1049	1048	1047	1046	1045	1044	1043	0	0	0	0	1	1	1	1	1
9·6	·1042	1041	1039	1038	1037	1036	1035	1034	1033	1032	0	0	0	0	1	1	1	1	1
9·7	·1031	1030	1029	1028	1027	1026	1025	1024	1022	1021	0	0	0	0	1	1	1	1	1
9·8	·1020	1019	1018	1017	1016	1015	1014	1013	1012	1011	0	0	0	0	1	1	1	1	1
9·9	·1010	1009	1008	1007	1006	1005	1004	1003	1002	1001	0	0	0	0	0	1	1	1	1

Square roots from 1 to 10

	0	1	2	3	4	5	6	7	8	9	Mean Differences 1 2 3	4 5 6	7 8 9
1·0	1·000	1·005	1·010	1·015	1·020	1·025	1·030	1·034	1·039	1·044	0 1 1	2 2 3	3 4 4
1·1	1·049	1·054	1·058	1·063	1·068	1·072	1·077	1·082	1·086	1·091	0 1 1	2 2 3	3 4 4
1·2	1·095	1·100	1·105	1·109	1·114	1·118	1·122	1·127	1·131	1·136	0 1 1	2 2 3	3 4 4
1·3	1·140	1·145	1·149	1·153	1·158	1·162	1·166	1·170	1·175	1·179	0 1 1	2 2 3	3 3 4
1·4	1·183	1·187	1·192	1·196	1·200	1·204	1·208	1·212	1·217	1·221	0 1 1	2 2 2	3 3 4
1·5	1·225	1·229	1·233	1·237	1·241	1·245	1·249	1·253	1·257	1·261	0 1 1	2 2 2	3 3 4
1·6	1·265	1·269	1·273	1·277	1·281	1·285	1·288	1·292	1·296	1·300	0 1 1	2 2 2	3 3 3
1·7	1·304	1·308	1·311	1·315	1·319	1·323	1·327	1·330	1·334	1·338	0 1 1	2 2 2	3 3 3
1·8	1·342	1·345	1·349	1·353	1·356	1·360	1·364	1·367	1·371	1·375	0 1 1	1 2 2	3 3 3
1·9	1·378	1·382	1·386	1·389	1·393	1·396	1·400	1·404	1·407	1·411	0 1 1	1 2 2	3 3 3
2·0	1·414	1·418	1·421	1·425	1·428	1·432	1·435	1·439	1·442	1·446	0 1 1	1 2 2	2 3 3
2·1	1·449	1·453	1·456	1·459	1·463	1·466	1·470	1·473	1·476	1·480	0 1 1	1 2 2	2 3 3
2·2	1·483	1·487	1·490	1·493	1·497	1·500	1·503	1·507	1·510	1·513	0 1 1	1 2 2	2 3 3
2·3	1·517	1·520	1·523	1·526	1·530	1·533	1·536	1·539	1·543	1·546	0 1 1	1 2 2	2 3 3
2·4	1·549	1·552	1·556	1·559	1·562	1·565	1·568	1·572	1·575	1·578	0 1 1	1 2 2	2 3 3
2·5	1·581	1·584	1·587	1·591	1·594	1·597	1·600	1·603	1·606	1·609	0 1 1	1 2 2	2 3 3
2·6	1·612	1·616	1·619	1·622	1·625	1·628	1·631	1·634	1·637	1·640	0 1 1	1 2 2	2 2 3
2·7	1·643	1·646	1·649	1·652	1·655	1·658	1·661	1·664	1·667	1·670	0 1 1	1 2 2	2 2 3
2·8	1·673	1·676	1·679	1·682	1·685	1·688	1·691	1·694	1·697	1·700	0 1 1	1 1 2	2 2 3
2·9	1·703	1·706	1·709	1·712	1·715	1·718	1·720	1·723	1·726	1·729	0 1 1	1 1 2	2 2 3
3·0	1·732	1·735	1·738	1·741	1·744	1·746	1·749	1·752	1·755	1·758	0 1 1	1 1 2	2 2 3
3·1	1·761	1·764	1·766	1·769	1·772	1·775	1·778	1·780	1·783	1·786	0 1 1	1 1 2	2 2 3
3·2	1·789	1·792	1·794	1·797	1·800	1·803	1·806	1·808	1·811	1·814	0 1 1	1 1 2	2 2 2
3·3	1·817	1·819	1·822	1·825	1·828	1·830	1·833	1·836	1·838	1·841	0 1 1	1 1 2	2 2 2
3·4	1·844	1·847	1·849	1·852	1·855	1·857	1·860	1·863	1·865	1·868	0 1 1	1 1 2	2 2 2
3·5	1·871	1·873	1·876	1·879	1·881	1·884	1·887	1·889	1·892	1·895	0 1 1	1 1 2	2 2 2
3·6	1·897	1·900	1·903	1·905	1·908	1·910	1·913	1·916	1·918	1·921	0 1 1	1 1 2	2 2 2
3·7	1·924	1·926	1·929	1·931	1·934	1·936	1·939	1·942	1·944	1·947	0 1 1	1 1 2	2 2 2
3·8	1·949	1·952	1·954	1·957	1·960	1·962	1·965	1·967	1·970	1·972	0 1 1	1 1 2	2 2 2
3·9	1·975	1·977	1·980	1·982	1·985	1·987	1·990	1·992	1·995	1·997	0 1 1	1 1 2	2 2 2
4·0	2·000	2·002	2·005	2·007	2·010	2·012	2·015	2·017	2·020	2·022	0 0 1	1 1 1	2 2 2
4·1	2·025	2·027	2·030	2·032	2·035	2·037	2·040	2·042	2·045	2·047	0 0 1	1 1 1	2 2 2
4·2	2·049	2·052	2·054	2·056	2·059	2·062	2·064	2·066	2·069	2·071	0 0 1	1 1 1	2 2 2
4·3	2·074	2·076	2·078	2·081	2·083	2·086	2·088	2·090	2·093	2·095	0 0 1	1 1 1	2 2 2
4·4	2·098	2·100	2·102	2·105	2·107	2·110	2·112	2·114	2·117	2·119	0 0 1	1 1 1	2 2 2
4·5	2·121	2·124	2·126	2·128	2·131	2·133	2·135	2·138	2·140	2·142	0 0 1	1 1 1	2 2 2
4·6	2·145	2·147	2·149	2·152	2·154	2·156	2·159	2·161	2·163	2·166	0 0 1	1 1 1	2 2 2
4·7	2·168	2·170	2·173	2·175	2·177	2·179	2·182	2·184	2·186	2·189	0 0 1	1 1 1	2 2 2
4·8	2·191	2·193	2·195	2·198	2·200	2·202	2·205	2·207	2·209	2·211	0 0 1	1 1 1	2 2 2
4·9	2·214	2·216	2·218	2·220	2·223	2·225	2·227	2·229	2·232	2·234	0 0 1	1 1 1	2 2 2
5·0	2·236	2·238	2·241	2·243	2·245	2·247	2·249	2·252	2·254	2·256	0 0 1	1 1 1	2 2 2
5·1	2·258	2·261	2·263	2·265	2·267	2·269	2·272	2·274	2·276	2·278	0 0 1	1 1 1	2 2 2
5·2	2·280	2·283	2·285	2·287	2·289	2·291	2·293	2·296	2·298	2·300	0 0 1	1 1 1	2 2 2
5·3	2·302	2·304	2·307	2·309	2·311	2·313	2·315	2·317	2·319	2·322	0 0 1	1 1 1	2 2 2
5·4	2·324	2·326	2·328	2·330	2·332	2·335	2·337	2·339	2·341	2·343	0 0 1	1 1 1	1 2 2

Square roots from 1 to 10

	0	1	2	3	4	5	6	7	8	9	Mean Differences.								
											1	2	3	4	5	6	7	8	9
5·5	2·345	2·347	2·349	2·352	2·354	2·356	2·358	2·360	2·362	2·364	0	0	1	1	1	1	1	2	2
5·6	2·366	2·369	2·371	2·373	2·375	2·377	2·379	2·381	2·383	2·385	0	0	1	1	1	1	1	2	2
5·7	2·387	2·390	2·392	2·394	2·396	2·398	2·400	2·402	2·404	2·406	0	0	1	1	1	1	1	2	2
5·8	2·408	2·410	2·412	2·415	2·417	2·419	2·421	2·423	2·425	2·427	0	0	1	1	1	1	1	2	2
5·9	2·429	2·431	2·433	2·435	2·437	2·439	2·441	2·443	2·445	2·447	0	0	1	1	1	1	1	2	2
6·0	2·449	2·452	2·454	2·456	2·458	2·460	2·462	2·464	2·466	2·468	0	0	1	1	1	1	1	2	2
6·1	2·470	2·472	2·474	2·476	2·478	2·480	2·482	2·484	2·486	2·488	0	0	1	1	1	1	1	2	2
6·2	2·4900	2·492	2·494	2·496	2·498	2·500	2·502	2·504	2·506	2·508	0	0	1	1	1	1	1	2	2
6·3	2·510	2·512	2·514	2·516	2·518	2·520	2·522	2·524	2·526	2·528	0	0	1	1	1	1	1	2	2
6·4	2·530	2·532	2·534	2·536	2·538	2·540	2·542	2·544	2·546	2·548	0	0	1	1	1	1	1	2	2
6·5	2·550	2·551	2·553	2·555	2·557	2·559	2·561	2·563	2·565	2·567	0	0	1	1	1	1	1	2	2
6·6	2·569	2·571	2·573	2·575	2·577	2·579	2·581	2·583	2·585	2·587	0	0	1	1	1	1	1	2	2
6·7	2·588	2·590	2·592	2·594	2·596	2·598	2·600	2·602	2·604	2·606	0	0	1	1	1	1	1	2	2
6·8	2·608	2·610	2·612	2·613	2·615	2·617	2·619	2·621	2·623	2·625	0	0	1	1	1	1	1	2	2
6·9	2·627	2·629	2·631	2·632	2·634	2·636	2·638	2·640	2·642	2·644	0	0	1	1	1	1	1	2	2
7·0	2·646	2·648	2·650	2·651	2·653	2·655	2·657	2·659	2·661	2·663	0	0	1	1	1	1	1	2	2
7·1	2·665	2·666	2·668	2·670	2·672	2·674	2·676	2·678	2·680	2·681	0	0	1	1	1	1	1	1	2
7·2	2·683	2·685	2·687	2·689	2·691	2·693	2·694	2·696	2·698	2·700	0	0	1	1	1	1	1	1	2
7·3	2·702	2·704	2·706	2·707	2·709	2·711	2·713	2·715	2·717	2·718	0	0	1	1	1	1	1	1	2
7·4	2·720	2·722	2·724	2·726	2·728	2·729	2·731	2·733	2·735	2·737	0	0	1	1	1	1	1	1	2
7·5	2·739	2·740	2·742	2·744	2·746	2·748	2·750	2·751	2·753	2·755	0	0	1	1	1	1	1	1	2
7·6	2·757	2·759	2·760	2·762	2·764	2·766	2·768	2·769	2·771	2·773	0	0	1	1	1	1	1	1	2
7·7	2·775	2·777	2·778	2·780	2·782	2·784	2·786	2·787	2·789	2·791	0	0	1	1	1	1	1	1	2
7·8	2·793	2·795	2·796	2·798	2·800	2·802	2·804	2·805	2·807	2·809	0	0	1	1	1	1	1	1	2
7·9	2·811	2·812	2·814	2·816	2·818	2·820	2·821	2·823	2·825	2·827	0	0	1	1	1	1	1	1	2
8·0	2·828	2·830	2·832	2·834	2·835	2·837	2·839	2·841	2·843	2·844	0	0	1	1	1	1	1	1	2
8·1	2·846	2·848	2·850	2·851	2·853	2·855	2·857	2·858	2·860	2·862	0	0	1	1	1	1	1	1	2
8·2	2·864	2·865	2·867	2·869	2·871	2·872	2·874	2·876	2·877	2·879	0	0	1	1	1	1	1	1	2
8·3	2·881	2·883	2·884	2·886	2·888	2·890	2·891	2·893	2·895	2·897	0	0	1	1	1	1	1	1	2
8·4	2·898	2·900	2·902	2·903	2·905	2·907	2·909	2·910	2·912	2·914	0	0	1	1	1	1	1	1	2
8·5	2·915	2·917	2·919	2·921	2·922	2·924	2·926	2·927	2·929	2·931	0	0	1	1	1	1	1	1	2
8·6	2·933	2·934	2·936	2·938	2·939	2·941	2·943	2·944	2·946	2·948	0	0	1	1	1	1	1	1	2
8·7	2·950	2·951	2·953	2·955	2·956	2·958	2·960	2·961	2·963	2·965	0	0	1	1	1	1	1	1	2
8·8	2·966	2·968	2·970	2·972	2·973	2·975	2·977	2·978	2·980	2·982	0	0	1	1	1	1	1	1	2
8·9	2·983	2·985	2·987	2·988	2·990	2·992	2·993	2·995	2·997	2·998	0	0	1	1	1	1	1	1	2
9·0	3·000	3·002	3·003	3·005	3·007	3·008	3·010	3·012	3·013	3·015	0	0	0	1	1	1	1	1	1
9·1	3·017	3·018	3·020	3·022	3·023	3·025	3·027	3·028	3·030	3·032	0	0	0	1	1	1	1	1	1
9·2	3·033	3·035	3·036	3·038	3·040	3·041	3·043	3·045	3·046	3·048	0	0	0	1	1	1	1	1	1
9·3	3·050	3·051	3·053	3·055	3·056	3·058	3·059	3·061	3·063	3·064	0	0	0	1	1	1	1	1	1
9·4	3·066	3·068	3·069	3·071	3·072	3·074	3·076	3·077	3·079	3·081	0	0	0	1	1	1	1	1	1
9·5	3·082	3·084	3·085	3·087	3·089	3·090	3·092	3·094	3·095	3·097	0	0	0	1	1	1	1	1	1
9·6	3·098	3·100	3·102	3·103	3·105	3·106	3·108	3·110	3·111	3·113	0	0	0	1	1	1	1	1	1
9·7	3·114	3·116	3·118	3·119	3·121	3·122	3·124	3·126	3·127	3·129	0	0	0	1	1	1	1	1	1
9·8	3·130	3·132	3·134	3·135	3·137	3·138	3·140	3·142	3·143	3·145	0	0	0	1	1	1	1	1	1
9·9	3·146	3·148	3·150	3·151	3·153	3·154	3·156	3·158	3·159	3·161	0	0	0	1	1	1	1	1	1

Square roots from 10 to 100

	0	1	2	3	4	5	6	7	8	9	Mean Differences								
											1	2	3	4	5	6	7	8	9
10	3·162	3·178	3·194	3·209	3·225	3·240	3·256	3·271	3·286	3·302	2	3	5	6	8	9	11	12	14
11	3·317	3·332	3·347	3·362	3·376	3·391	3·406	3·421	3·435	3·450	1	3	4	6	7	9	10	12	13
12	3·464	3·479	3·493	3·507	3·521	3·536	3·550	3·564	3·578	3·592	1	3	4	6	7	8	10	11	13
13	3·606	3·619	3·633	3·647	3·661	3·674	3·688	3·701	3·715	3·728	1	3	4	5	7	8	10	11	12
14	3·742	3·755	3·768	3·782	3·795	3·808	3·821	3·834	3·847	3·860	1	3	4	5	7	8	9	11	12
15	3·873	3·886	3·899	3·912	3·924	3·937	3·950	3·962	3·975	3·987	1	3	4	5	6	8	9	10	11
16	4·000	4·012	4·025	4·037	4·050	4·062	4·074	4·087	4·099	4·111	1	2	4	5	6	7	9	10	11
17	4·123	4·135	4·147	4·159	4·171	4·183	4·195	4·207	4·219	4·231	1	2	4	5	6	7	8	10	11
18	4·243	4·254	4·266	4·278	4·290	4·301	4·313	4·324	4·336	4·347	1	2	3	5	6	7	8	9	10
19	4·359	4·370	4·382	4·393	4·405	4·416	4·427	4·438	4·450	4·461	1	2	3	5	6	7	8	9	10
20	4·472	4·483	4·494	4·506	4·517	4·528	4·539	4·550	4·561	4·572	1	2	3	4	6	7	8	9	10
21	4·583	4·593	4·604	4·615	4·626	4·637	4·648	4·658	4·669	4·680	1	2	3	4	5	6	8	9	10
22	4·690	4·701	4·712	4·722	4·733	4·743	4·754	4·764	4·775	4·785	1	2	3	4	5	6	7	8	9
23	4·796	4·806	4·817	4·827	4·837	4·848	4·858	4·868	4·879	4·889	1	2	3	4	5	6	7	8	9
24	4·899	4·909	4·919	4·930	4·940	4·950	4·960	4·970	4·980	4·990	1	2	3	4	5	6	7	8	9
25	5·000	5·010	5·020	5·030	5·040	5·050	5·060	5·070	5·079	5·089	1	2	3	4	5	6	7	8	9
26	5·099	5·109	5·119	5·128	5·138	5·148	5·158	5·167	5·177	5·187	1	2	3	4	5	6	7	8	9
27	5·196	5·206	5·215	5·225	5·235	5·244	5·254	5·263	5·273	5·282	1	2	3	4	5	6	7	8	9
28	5·292	5·301	5·310	5·320	5·329	5·339	5·348	5·357	5·367	5·376	1	2	3	4	5	6	7	7	8
29	5·385	5·394	5·404	5·413	5·422	5·431	5·441	5·450	5·459	5·468	1	2	3	4	5	5	6	7	8
30	5·477	5·486	5·495	5·505	5·514	5·523	5·532	5·541	5·550	5·559	1	2	3	4	4	5	6	7	8
31	5·568	5·577	5·586	5·595	5·604	5·612	5·621	5·630	5·639	5·648	1	2	3	3	4	5	6	7	8
32	5·657	5·666	5·675	5·683	5·692	5·701	5·710	5·718	5·727	5·736	1	2	3	3	4	5	6	7	8
33	5·745	5·753	5·762	5·771	5·779	5·788	5·797	5·805	5·814	5·822	1	2	3	3	4	5	6	7	8
34	5·831	5·840	5·848	5·857	5·865	5·874	5·882	5·891	5·899	5·908	1	2	3	3	4	5	6	7	8
35	5·916	5·925	5·933	5·941	5·950	5·958	5·967	5·975	5·983	5·992	1	2	2	3	4	5	6	7	8
36	6·000	6·008	6·017	6·025	6·033	6·042	6·050	6·058	6·066	6·075	1	2	2	3	4	5	6	7	7
37	6·083	6·091	6·099	6·107	6·116	6·124	6·132	6·140	6·148	6·156	1	2	2	3	4	5	6	7	7
38	6·164	6·173	6·181	6·189	6·197	6·205	6·213	6·221	6·229	6·237	1	2	2	3	4	5	6	6	7
39	6·245	6·253	6·261	6·269	6·277	6·285	6·293	6·301	6·309	6·317	1	2	2	3	4	5	6	6	7
40	6·325	6·332	6·340	6·348	6·356	6·364	6·372	6·380	6·387	6·395	1	2	2	3	4	5	6	6	7
41	6·403	6·411	6·419	6·427	6·434	6·442	6·450	6·458	6·465	6·473	1	2	2	3	4	5	5	6	7
42	6·481	6·488	6·496	6·504	6·512	6·519	6·527	6·535	6·542	6·550	1	2	2	3	4	5	5	6	7
43	6·557	6·565	6·573	6·580	6·588	6·595	6·603	6·611	6·618	6·626	1	2	2	3	4	5	5	6	7
44	6·633	6·641	6·648	6·656	6·663	6·671	6·678	6·686	6·693	6·701	1	2	2	3	4	5	5	6	7
45	6·708	6·716	6·723	6·731	6·738	6·745	6·753	6·760	6·768	6·775	1	1	2	3	4	4	5	6	7
46	6·782	6·790	6·797	6·804	6·812	6·819	6·826	6·834	6·841	6·848	1	1	2	3	4	4	5	6	7
47	6·856	6·863	6·870	6·877	6·885	6·892	6·899	6·907	6·914	6·921	1	1	2	3	4	4	5	6	7
48	6·928	6·935	6·943	6·950	6·957	6·964	6·971	6·979	6·986	6·993	1	1	2	3	4	4	5	6	6
49	7·000	7·007	7·014	7·021	7·029	7·036	7·043	7·050	7·057	7·064	1	1	2	3	4	4	5	6	6
50	7·071	7·078	7·085	7·092	7·099	7·106	7·113	7·120	7·127	7·134	1	1	2	3	4	4	5	6	6
51	7·141	7·148	7·155	7·162	7·169	7·176	7·183	7·190	7·197	7·204	1	1	2	3	4	4	5	6	6
52	7·211	7·218	7·225	7·232	7·239	7·246	7·253	7·259	7·266	7·273	1	1	2	3	3	4	5	6	6
53	7·280	7·287	7·294	7·301	7·308	7·314	7·321	7·328	7·335	7·342	1	1	2	3	3	4	5	5	6
54	7·348	7·355	7·362	7·369	7·376	7·382	7·389	7·396	7·403	7·409	1	1	2	3	3	4	5	5	6

Square roots from 10 to 100

	0	1	2	3	4	5	6	7	8	9	Mean Differences 1 2 3	4 5 6	7 8 9
55	7·416	7·423	7·430	7·436	7·443	7·450	7·457	7·463	7·470	7·477	1 1 2	3 3 4	5 5 6
56	7·483	7·490	7·497	7·503	7·510	7·517	7·523	7·530	7·537	7·543	1 1 2	3 3 4	5 5 6
57	7·550	7·556	7·563	7·570	7·576	7·583	7·589	7·596	7·603	7·609	1 1 2	3 3 4	5 5 6
58	7·616	7·622	7·629	7·635	7·642	7·649	7·655	7·662	7·668	7·675	1 1 2	3 3 4	5 5 6
59	7·681	7·688	7·694	7·701	7·707	7·714	7·720	7·727	7·733	7·740	1 1 2	3 3 4	4 5 6
60	7·746	7·752	7·759	7·765	7·772	7·778	7·785	7·791	7·797	7·804	1 1 2	3 3 4	4 5 6
61	7·810	7·817	7·823	7·829	7·836	7·842	7·849	7·855	7·861	7·868	1 1 2	3 3 4	4 5 6
62	7·874	7·880	7·887	7·893	7·899	7·906	7·912	7·918	7·925	7·931	1 1 2	3 3 4	4 5 6
63	7·937	7·944	7·950	7·956	7·962	7·969	7·975	7·981	7·987	7·994	1 1 2	3 3 4	4 5 6
64	8·000	8·006	8·012	8·019	8·025	8·031	8·037	8·044	8·050	8·056	1 1 2	2 3 4	4 5 6
65	8·062	8·068	8·075	8·081	8·087	8·093	8·099	8·106	8·112	8·118	1 1 2	2 3 4	4 5 6
66	8·124	8·130	8·136	8·142	8·149	8·155	8·161	8·167	8·173	8·179	1 1 2	2 3 4	4 5 5
67	8·185	8·191	8·198	8·204	8·210	8·216	8·222	8·228	8·234	8·240	1 1 2	2 3 4	4 5 5
68	8·246	8·252	8·258	8·264	8·270	8·276	8·283	8·289	8·295	8·301	1 1 2	2 3 4	4 5 5
69	8·307	8·313	8·319	8·325	8·331	8·337	8·343	8·349	8·355	8·361	1 1 2	2 3 4	4 5 5
70	8·367	8·373	8·379	8·385	8·390	8·396	8·402	8·408	8·414	8·420	1 1 2	2 3 4	4 5 5
71	8·426	8·432	8·438	8·444	8·450	8·456	8·462	8·468	8·473	8·479	1 1 2	2 3 4	4 5 5
72	8·485	8·491	8·497	8·503	8·509	8·515	8·521	8·526	8·532	8·538	1 1 2	2 3 3	4 5 5
73	8·544	8·550	8·556	8·562	8·567	8·573	8·579	8·585	8·591	8·597	1 1 2	2 3 3	4 5 5
74	8·602	8·608	8·614	8·620	8·626	8·631	8·637	8·643	8·649	8·654	1 1 2	2 3 3	4 5 5
75	8·660	8·666	8·672	8·678	8·683	8·689	8·695	8·701	8·706	8·712	1 1 2	2 3 3	4 5 5
76	8·718	8·724	8·729	8·735	8·741	8·746	8·752	8·758	8·764	8·769	1 1 2	2 3 3	4 5 5
77	8·775	8·781	8·786	8·792	8·798	8·803	8·809	8·815	8·820	8·826	1 1 2	2 3 3	4 4 5
78	8·832	8·837	8·843	8·849	8·854	8·860	8·866	8·871	8·877	8·883	1 1 2	2 3 3	4 4 5
79	8·888	8·894	8·899	8·905	8·911	8·916	8·922	8·927	8·933	8·939	1 1 2	2 3 3	4 4 5
80	8·944	8·950	8·955	8·961	8·967	8·972	8·978	8·983	8·989	8·994	1 1 2	2 3 3	4 4 5
81	9·000	9·006	9·011	9·017	9·022	9·028	9·033	9·039	9·044	9·050	1 1 2	2 3 3	4 4 5
82	9·055	9·061	9·066	9·072	9·077	9·083	9·088	9·094	9·099	9·105	1 1 2	2 3 3	4 4 5
83	9·110	9·116	9·121	9·127	9·132	9·138	9·143	9·149	9·154	9·160	1 1 2	2 3 3	4 4 5
84	9·165	9·171	9·176	9·182	9·187	9·192	9·198	9·203	9·209	9·214	1 1 2	2 3 3	4 4 5
85	9·220	9·225	9·230	9·236	9·241	9·247	9·252	9·257	9·263	9·268	1 1 2	2 3 3	4 4 5
86	9·274	9·279	9·284	9·290	9·295	9·301	9·306	9·311	9·317	9·322	1 1 2	2 3 3	4 4 5
87	9·327	9·333	9·338	9·343	9·349	9·354	9·359	9·365	9·370	9·375	1 1 2	2 3 3	4 4 5
88	9·381	9·386	9·391	9·397	9·402	9·407	9·413	9·418	9·423	9·429	1 1 2	2 3 3	4 4 5
89	9·434	9·439	9·445	9·450	9·455	9·460	9·466	9·471	9·476	9·482	1 1 2	2 3 3	4 4 5
90	9·487	9·492	9·497	9·503	9·508	9·513	9·518	9·524	9·529	9·534	1 1 2	2 3 3	4 4 5
91	9·539	9·545	9·550	9·555	9·560	9·566	9·571	9·576	9·581	9·586	1 1 2	2 3 3	4 4 5
92	9·592	9·597	9·602	9·607	9·612	9·618	9·623	9·628	9·633	9·638	1 1 2	2 3 3	4 4 5
93	9·644	9·649	9·654	9·659	9·664	9·670	9·675	9·680	9·685	9·690	1 1 2	2 3 3	4 4 5
94	9·695	9·701	9·706	9·711	9·716	9·721	9·726	9·731	9·737	9·742	1 1 2	2 3 3	4 4 5
95	9·747	9·752	9·757	9·762	9·767	9·772	9·778	9·783	9·788	9·793	1 1 2	2 3 3	4 4 5
96	9·798	9·803	9·808	9·813	9·818	9·823	9·829	9·834	9·839	9·844	1 1 2	2 3 3	4 4 5
97	9·849	9·854	9·859	9·864	9·869	9·874	9·879	9·884	9·889	9·894	1 1 1	2 3 3	4 4 5
98	9·899	9·905	9·910	9·915	9·920	9·925	9·930	9·935	9·940	9·945	0 1 1	2 2 3	3 4 4
99	9·950	9·955	9·960	9·965	9·970	9·975	9·980	9·985	9·990	9·995	0 1 1	2 2 3	3 4 4

Logarithms

	0	1	2	3	4	5	6	7	8	9	1 2 3	4 5 6	7 8 9
10	0000	0043	0086	0128	0170						5 9 13	17 21 26	30 34 38
						0212	0253	0294	0334	0374	4 8 12	16 20 24	28 32 36
11	0414	0453	0492	0531	0569						4 8 12	16 20 23	27 31 35
						0607	0645	0682	0719	0755	4 7 11	15 18 22	26 29 33
12	0792	0828	0864	0899	0934						3 7 11	14 18 21	25 28 32
						0969	1004	1038	1072	1106	3 7 10	14 17 20	24 27 31
13	1139	1173	1206	1239	1271						3 6 10	13 16 19	23 26 29
						1303	1335	1367	1399	1430	3 7 10	13 16 19	22 25 29
14	1461	1492	1523	1553	1584						3 6 9	12 15 19	22 25 28
						1614	1644	1673	1703	1732	3 6 9	12 14 17	20 23 26
15	1761	1790	1818	1847	1875						3 6 9	11 14 17	20 23 26
						1903	1931	1959	1987	2014	3 6 8	11 14 17	19 22 25
16	2041	2068	2095	2122	2148						3 6 8	11 14 16	19 22 24
						2175	2201	2227	2253	2279	3 5 8	10 13 16	18 21 23
17	2304	2330	2355	2380	2405						3 5 8	10 13 15	18 20 23
						2430	2455	2480	2504	2529	3 5 8	10 12 15	17 20 22
18	2553	2577	2601	2625	2648						2 5 7	9 12 14	17 19 21
						2672	2695	2718	2742	2765	2 4 7	9 11 14	16 18 21
19	2788	2810	2833	2856	2878						2 4 7	9 11 13	16 18 20
						2900	2923	2945	2967	2989	2 4 6	8 11 13	15 17 19
20	3010	3032	3054	3075	3096	3118	3139	3160	3181	3201	2 4 6	8 11 13	15 17 19
21	3222	3243	3263	3284	3304	3324	3345	3365	3385	3404	2 4 6	8 10 12	14 16 18
22	3424	3444	3464	3483	3502	3522	3541	3560	3579	3598	2 4 6	8 10 12	14 15 17
23	3617	3636	3655	3674	3692	3711	3729	3747	3766	3784	2 4 6	7 9 11	13 15 17
24	3802	3820	3838	3856	3874	3892	3909	3927	3945	3962	2 4 5	7 9 11	12 14 16
25	3979	3997	4014	4031	4048	4065	4082	4099	4116	4133	2 3 5	7 9 10	12 14 15
26	4150	4166	4183	4200	4216	4232	4249	4265	4281	4298	2 3 5	7 8 10	11 13 15
27	4314	4330	4346	4362	4378	4393	4409	4425	4440	4456	2 3 5	6 8 9	11 13 14
28	4472	4487	4502	4518	4533	4548	4564	4579	4594	4609	2 3 5	6 8 9	11 12 14
29	4624	4639	4654	4669	4683	4698	4713	4728	4742	4757	1 3 4	6 7 9	10 12 13
30	4771	4786	4800	4814	4829	4843	4857	4871	4886	4900	1 3 4	6 7 9	10 11 13
31	4914	4928	4942	4955	4969	4983	4997	5011	5024	5038	1 3 4	6 7 8	10 11 12
32	5051	5065	5079	5092	5105	5119	5132	5145	5159	5172	1 3 4	5 7 8	9 11 12
33	5185	5198	5211	5224	5237	5250	5263	5276	5289	5302	1 3 4	5 6 8	9 10 12
34	5315	5328	5340	5353	5366	5378	5391	5403	5416	5428	1 3 4	5 6 8	9 10 11
35	5441	5453	5465	5478	5490	5502	5514	5527	5539	5551	1 2 4	5 6 7	9 10 11
36	5563	5575	5587	5599	5611	5623	5635	5647	5658	5670	1 2 4	5 6 7	8 10 11
37	5682	5694	5705	5717	5729	5740	5752	5763	5775	5786	1 2 3	5 6 7	8 9 10
38	5798	5809	5821	5832	5843	5855	5866	5877	5888	5899	1 2 3	5 6 7	8 9 10
39	5911	5922	5933	5944	5955	5966	5977	5988	5999	6010	1 2 3	4 5 7	8 9 10
40	6021	6031	6042	6053	6064	6075	6085	6096	6107	6117	1 2 3	4 5 6	8 9 10
41	6128	6138	6149	6160	6170	6180	6191	6201	6212	6222	1 2 3	4 5 6	7 8 9
42	6232	6243	6253	6263	6274	6284	6294	6304	6314	6325	1 2 3	4 5 6	7 8 9
43	6335	6345	6355	6365	6375	6385	6395	6405	6415	6425	1 2 3	4 5 6	7 8 9
44	6435	6444	6454	6464	6474	6484	6493	6503	6513	6522	1 2 3	4 5 6	7 8 9
45	6532	6542	6551	6561	6571	6580	6590	6599	6609	6618	1 2 3	4 5 6	7 8 9
46	6628	6637	6646	6656	6665	6675	6684	6693	6702	6712	1 2 3	4 5 6	7 7 8
47	6721	6730	6739	6749	6758	6767	6776	6785	6794	6803	1 2 3	4 5 5	6 7 8
48	6812	6821	6830	6839	6848	6857	6866	6875	6884	6893	1 2 3	4 4 5	6 7 8
49	6902	6911	6920	6928	6937	6946	6955	6964	6972	6981	1 2 3	4 4 5	6 7 8

Logarithms

	0	1	2	3	4	5	6	7	8	9	123	456	789
50	6990	6998	7007	7016	7024	7033	7042	7050	7059	7067	1 2 3	3 4 5	6 7 8
51	7076	7084	7093	7101	7110	7118	7126	7135	7143	7152	1 2 3	3 4 5	6 7 8
52	7160	7168	7177	7185	7193	7202	7210	7218	7226	7235	1 2 2	3 4 5	6 7 7
53	7243	7251	7259	7267	7275	7284	7292	7300	7308	7316	1 2 2	3 4 5	6 6 7
54	7324	7332	7340	7348	7356	7364	7372	7380	7388	7396	1 2 2	3 4 5	6 6 7
55	7404	7412	7419	7427	7435	7443	7451	7459	7466	7474	1 2 2	3 4 5	5 6 7
56	7482	7490	7497	7505	7513	7520	7528	7536	7543	7551	1 2 2	3 4 5	5 6 7
57	7559	7566	7574	7582	7589	7597	7604	7612	7619	7627	1 2 2	3 4 5	5 6 7
58	7634	7642	7649	7657	7664	7672	7679	7686	7694	7701	1 1 2	3 4 4	5 6 7
59	7709	7716	7723	7731	7738	7745	7752	7760	7767	7774	1 1 2	3 4 4	5 6 7
60	7782	7789	7796	7803	7810	7818	7825	7832	7839	7846	1 1 2	3 4 4	5 6 6
61	7853	7860	7868	7875	7882	7889	7896	7903	7910	7917	1 1 2	3 4 4	5 6 6
62	7924	7931	7938	7945	7952	7959	7966	7973	7980	7987	1 1 2	3 3 4	5 6 6
63	7993	8000	8007	8014	8021	8028	8035	8041	8048	8055	1 1 2	3 3 4	5 5 6
64	8062	8069	8075	8082	8089	8096	8102	8109	8116	8122	1 1 2	3 3 4	5 5 6
65	8129	8136	8142	8149	8156	8162	8169	8176	8182	8189	1 1 2	3 3 4	5 5 6
66	8195	8202	8209	8215	8222	8228	8235	8241	8248	8254	1 1 2	3 3 4	5 5 6
67	8261	8267	8274	8280	8287	8293	8299	8306	8312	8319	1 1 2	3 3 4	5 5 6
68	8325	8331	8338	8344	8351	8357	8363	8370	8376	8382	1 1 2	3 3 4	4 5 6
69	8388	8395	8401	8407	8414	8420	8426	8432	8439	8445	1 1 2	2 3 4	4 5 6
70	8451	8457	8463	8470	8476	8482	8488	8494	8500	8506	1 1 2	2 3 4	4 5 6
71	8513	8519	8525	8531	8537	8543	8549	8555	8561	8567	1 1 2	2 3 4	4 5 5
72	8573	8579	8585	8591	8597	8603	8609	8615	8621	8627	1 1 2	2 3 4	4 5 5
73	8633	8639	8645	8651	8657	8663	8669	8675	8681	8686	1 1 2	2 3 4	4 5 5
74	8692	8698	8704	8710	8716	8722	8727	8733	8739	8745	1 1 2	2 3 4	4 5 5
75	8751	8756	8762	8768	8774	8779	8785	8791	8797	8802	1 1 2	2 3 3	4 5 5
76	8808	8814	8820	8825	8831	8837	8842	8848	8854	8859	1 1 2	2 3 3	4 5 5
77	8865	8871	8876	8882	8887	8893	8899	8904	8910	8915	1 1 2	2 3 3	4 4 5
78	8921	8927	8932	8938	8943	8949	8954	8960	8965	8971	1 1 2	2 3 3	4 4 5
79	8976	8982	8987	8993	8998	9004	9009	9015	9020	9025	1 1 2	2 3 3	4 4 5
80	9031	9036	9042	9047	9053	9058	9063	9069	9074	9079	1 1 2	2 3 3	4 4 5
81	9085	9090	9096	9101	9106	9112	9117	9122	9128	9133	1 1 2	2 3 3	4 4 5
82	9138	9143	9149	9154	9159	9165	9170	9175	9180	9186	1 1 2	2 3 3	4 4 5
83	9191	9196	9201	9206	9212	9217	9222	9227	9232	9238	1 1 2	2 3 3	4 4 5
84	9243	9248	9253	9258	9263	9269	9274	9279	9284	9289	1 1 2	2 3 3	4 4 5
85	9294	9299	9304	9309	9315	9320	9325	9330	9335	9340	1 1 2	2 3 3	4 4 5
86	9345	9350	9355	9360	9365	9370	9375	9380	9385	9390	1 1 2	2 3 3	4 4 5
87	9395	9400	9405	9410	9415	9420	9425	9430	9435	9440	0 1 1	2 2 3	3 4 4
88	9445	9450	9455	9460	9465	9469	9474	9479	9484	9489	0 1 1	2 2 3	3 4 4
89	9494	9499	9504	9509	9513	9518	9523	9528	9533	9538	0 1 1	2 2 3	3 4 4
90	9542	9547	9552	9557	9562	9566	9571	9576	9581	9586	0 1 1	2 2 3	3 4 4
91	9590	9595	9600	9605	9609	9614	9619	9624	9628	9633	0 1 1	2 2 3	3 4 4
92	9638	9643	9647	9652	9657	9661	9666	9671	9675	9680	0 1 1	2 2 3	3 4 4
93	9685	9689	9694	9699	9703	9708	9713	9717	9722	9727	0 1 1	2 2 3	3 4 4
94	9731	9736	9741	9745	9750	9754	9759	9763	9768	9773	0 1 1	2 2 3	3 4 4
95	9777	9782	9786	9791	9795	9800	9805	9809	9814	9818	0 1 1	2 2 3	3 4 4
96	9823	9827	9832	9836	9841	9845	9850	9854	9859	9863	0 1 1	2 2 3	3 4 4
97	9868	9872	9877	9881	9886	9890	9894	9899	9903	9908	0 1 1	2 2 3	3 4 4
98	9912	9917	9921	9926	9930	9934	9939	9943	9948	9952	0 1 1	2 2 3	3 4 4
99	9956	9961	9965	9969	9974	9978	9983	9987	9991	9996	0 1 1	2 2 3	3 3 4

Antilogarithms

	0	1	2	3	4	5	6	7	8	9	1 2 3	4 5 6	7 8 9
·00	1000	1002	1005	1007	1009	1012	1014	1016	1019	1021	0 0 1	1 1 1	2 2 2
·01	1023	1026	1028	1030	1033	1035	1038	1040	1042	1045	0 0 1	1 1 1	2 2 2
·02	1047	1050	1052	1054	1057	1059	1062	1064	1067	1069	0 0 1	1 1 1	2 2 2
·03	1072	1074	1076	1079	1081	1084	1086	1089	1091	1094	0 0 1	1 1 1	2 2 2
·04	1096	1099	1102	1104	1107	1109	1112	1114	1117	1119	0 1 1	1 1 2	2 2 2
·05	1122	1125	1127	1130	1132	1135	1138	1140	1143	1146	0 1 1	1 1 2	2 2 2
·06	1148	1151	1153	1156	1159	1161	1164	1167	1169	1172	0 1 1	1 1 2	2 2 2
·07	1175	1178	1180	1183	1186	1189	1191	1194	1197	1199	0 1 1	1 1 2	2 2 2
·08	1202	1205	1208	1211	1213	1216	1219	1222	1225	1227	0 1 1	1 1 2	2 2 3
·09	1230	1233	1236	1239	1242	1245	1247	1250	1253	1256	0 1 1	1 1 2	2 2 3
·10	1259	1262	1265	1268	1271	1274	1276	1279	1282	1285	0 1 1	1 1 2	2 2 3
·11	1288	1291	1294	1297	1300	1303	1306	1309	1312	1315	0 1 1	1 2 2	2 2 3
·12	1318	1321	1324	1327	1330	1334	1337	1340	1343	1346	0 1 1	1 2 2	2 2 3
·13	1349	1352	1355	1358	1361	1365	1368	1371	1374	1377	0 1 1	1 2 2	2 3 3
·14	1380	1384	1387	1390	1393	1396	1400	1403	1406	1409	0 1 1	1 2 2	2 3 3
·15	1413	1416	1419	1422	1426	1429	1432	1435	1439	1442	0 1 1	1 2 2	2 3 3
·16	1445	1449	1452	1455	1459	1462	1466	1469	1472	1476	0 1 1	1 2 2	2 3 3
·17	1479	1483	1486	1489	1493	1496	1500	1503	1507	1510	0 1 1	1 2 2	2 3 3
·18	1514	1517	1521	1524	1528	1531	1535	1538	1542	1545	0 1 1	1 2 2	2 3 3
·19	1549	1552	1556	1560	1563	1567	1570	1574	1578	1581	0 1 1	1 2 2	3 3 3
·20	1585	1589	1592	1596	1600	1603	1607	1611	1614	1618	0 1 1	1 2 2	3 3 3
·21	1622	1626	1629	1633	1637	1641	1644	1648	1652	1656	0 1 1	2 2 2	3 3 3
·22	1660	1663	1667	1671	1675	1679	1683	1687	1690	1694	0 1 1	2 2 2	3 3 3
·23	1698	1702	1706	1710	1714	1718	1722	1726	1730	1734	0 1 1	2 2 2	3 3 4
·24	1738	1742	1746	1750	1754	1758	1762	1766	1770	1774	0 1 1	2 2 2	3 3 4
·25	1778	1782	1786	1791	1795	1799	1803	1807	1811	1816	0 1 1	2 2 2	3 3 4
·26	1820	1824	1828	1832	1837	1841	1845	1849	1854	1858	0 1 1	2 2 3	3 3 4
·27	1862	1866	1871	1875	1879	1884	1888	1892	1897	1901	0 1 1	2 2 3	3 3 4
·28	1905	1910	1914	1919	1923	1928	1932	1936	1941	1945	0 1 1	2 2 3	3 4 4
·29	1950	1954	1959	1963	1968	1972	1977	1982	1986	1991	0 1 1	2 2 3	3 4 4
·30	1995	2000	2004	2009	2014	2018	2023	2028	2032	2037	0 1 1	2 2 3	3 4 4
·31	2042	2046	2051	2056	2061	2065	2070	2075	2080	2084	0 1 1	2 2 3	3 4 4
·32	2089	2094	2099	2104	2109	2113	2118	2123	2128	2133	0 1 1	2 2 3	3 4 4
·33	2138	2143	2148	2153	2158	2163	2168	2173	2178	2183	0 1 1	2 2 3	3 4 4
·34	2188	2193	2198	2203	2208	2213	2218	2223	2228	2234	1 1 2	2 3 3	4 4 5
·35	2239	2244	2249	2254	2259	2265	2270	2275	2280	2286	1 1 2	2 3 3	4 4 5
·36	2291	2296	2301	2307	2312	2317	2323	2328	2333	2339	1 1 2	2 3 3	4 4 5
·37	2344	2350	2355	2360	2366	2371	2377	2382	2388	2393	1 1 2	2 3 3	4 4 5
·38	2399	2404	2410	2415	2421	2427	2432	2438	2443	2449	1 1 2	2 3 3	4 4 5
·39	2455	2460	2466	2472	2477	2483	2489	2495	2500	2506	1 1 2	2 3 3	4 5 5
·40	2512	2518	2523	2529	2535	2541	2547	2553	2559	2564	1 1 2	2 3 4	4 5 5
·41	2570	2576	2582	2588	2594	2600	2606	2612	2618	2624	1 1 2	2 3 4	4 5 5
·42	2630	2636	2642	2649	2655	2661	2667	2673	2679	2685	1 1 2	2 3 4	4 5 6
·43	2692	2698	2704	2710	2716	2723	2729	2735	2742	2748	1 1 2	3 3 4	4 5 6
·44	2754	2761	2767	2773	2780	2786	2793	2799	2805	2812	1 1 2	3 3 4	4 5 6
·45	2818	2825	2831	2838	2844	2851	2858	2864	2871	2877	1 1 2	3 3 4	5 5 6
·46	2884	2891	2897	2904	2911	2917	2924	2931	2938	2944	1 1 2	3 3 4	5 5 6
·47	2951	2958	2965	2972	2979	2985	2992	2999	3006	3013	1 1 2	3 3 4	5 5 6
·48	3020	3027	3034	3041	3048	3055	3062	3069	3076	3083	1 1 2	3 4 4	5 6 6
·49	3090	3097	3105	3112	3119	3126	3133	3141	3148	3155	1 1 2	3 4 4	5 6 6

Antilogarithms

	0	1	2	3	4	5	6	7	8	9	1 2 3	4 5 6	7 8 9
·50	3162	3170	3177	3184	3192	3199	3206	3214	3221	3228	1 1 2	3 4 4	5 6 7
·51	3236	3243	3251	3258	3266	3273	3281	3289	3296	3304	1 2 2	3 4 5	5 6 7
·52	3311	3319	3327	3334	3342	3350	3357	3365	3373	3381	1 2 2	3 4 5	5 6 7
·53	3388	3396	3404	3412	3420	3428	3436	3443	3451	3459	1 2 2	3 4 5	6 6 7
·54	3467	3475	3483	3491	3499	3508	3516	3524	3532	3540	1 2 2	3 4 5	6 6 7
·55	3548	3556	3565	3573	3581	3589	3597	3606	3614	3622	1 2 2	3 4 5	6 7 7
·56	3631	3639	3648	3656	3664	3673	3681	3690	3698	3707	1 2 3	3 4 5	6 7 8
·57	3715	3724	3733	3741	3750	3758	3767	3776	3784	3793	1 2 3	3 4 5	6 7 8
·58	3802	3811	3819	3828	3837	3846	3855	3864	3873	3882	1 2 3	4 4 5	6 7 8
·59	3890	3899	3908	3917	3926	3936	3945	3954	3963	3972	1 2 3	4 5 5	6 7 8
·60	3981	3990	3999	4009	4018	4027	4036	4046	4055	4064	1 2 3	4 5 6	6 7 8
·61	4074	4083	4093	4102	4111	4121	4130	4140	4150	4159	1 2 3	4 5 6	7 8 9
·62	4169	4178	4188	4198	4207	4217	4227	4236	4246	4256	1 2 3	4 5 6	7 8 9
·63	4266	4276	4285	4295	4305	4315	4325	4335	4345	4355	1 2 3	4 5 6	7 8 9
·64	4365	4375	4385	4395	4406	4416	4426	4436	4446	4457	1 2 3	4 5 6	7 8 9
·65	4467	4477	4487	4498	4508	4519	4529	4539	4550	4560	1 2 3	4 5 6	7 8 9
·66	4571	4581	4592	4603	4613	4624	4634	4645	4656	4667	1 2 3	4 5 6	7 9 10
·67	4677	4688	4699	4710	4721	4732	4742	4753	4764	4775	1 2 3	4 5 7	8 9 10
·68	4786	4797	4808	4819	4831	4842	4853	4864	4875	4887	1 2 3	4 6 7	8 9 10
·69	4898	4909	4920	4932	4943	4955	4966	4977	4989	5000	1 2 3	5 6 7	8 9 10
·70	5012	5023	5035	5047	5058	5070	5082	5093	5105	5117	1 2 4	5 6 7	8 9 11
·71	5129	5140	5152	5164	5176	5188	5200	5212	5224	5236	1 2 4	5 6 7	8 10 11
·72	5248	5260	5272	5284	5297	5309	5321	5333	5346	5358	1 2 4	5 6 7	9 10 11
·73	5370	5383	5395	5408	5420	5433	5445	5458	5470	5483	1 3 4	5 6 8	9 10 11
·74	5495	5508	5521	5534	5546	5559	5572	5585	5598	5610	1 3 4	5 6 8	9 10 12
·75	5623	5636	5649	5662	5675	5689	5702	5715	5728	5741	1 3 4	5 7 8	9 10 12
·76	5754	5768	5781	5794	5808	5821	5834	5848	5861	5875	1 3 4	5 7 8	9 11 12
·77	5888	5902	5916	5929	5943	5957	5970	5984	5998	6012	1 3 4	5 7 8	10 11 12
·78	6026	6039	6053	6067	6081	6095	6109	6124	6138	6152	1 3 4	6 7 8	10 11 13
·79	6166	6180	6194	6209	6223	6237	6252	6266	6281	6295	1 3 4	6 7 9	10 11 13
·80	6310	6324	6339	6353	6368	6383	6397	6412	6427	6442	1 3 4	6 7 9	10 12 13
·81	6457	6471	6486	6501	6516	6531	6546	6561	6577	6592	2 3 5	6 8 9	11 12 14
·82	6607	6622	6637	6653	6668	6683	6699	6714	6730	6745	2 3 5	6 8 9	11 12 14
·83	6761	6776	6792	6808	6823	6839	6855	6871	6887	6902	2 3 5	6 8 9	11 13 14
·84	6918	6934	6950	6966	6982	6998	7015	7031	7047	7063	2 3 5	6 8 10	11 13 15
·85	7079	7096	7112	7129	7145	7161	7178	7194	7211	7228	2 3 5	7 8 10	12 13 15
·86	7244	7261	7278	7295	7311	7328	7345	7362	7379	7396	2 3 5	7 8 10	12 13 15
·87	7413	7430	7447	7464	7482	7499	7516	7534	7551	7568	2 3 5	7 9 10	12 14 16
·88	7586	7603	7621	7638	7656	7674	7691	7709	7727	7745	2 4 5	7 9 11	12 14 16
·89	7762	7780	7798	7816	7834	7852	7870	7889	7907	7925	2 4 5	7 9 11	13 14 16
·90	7943	7962	7980	7998	8017	8035	8054	8072	8091	8110	2 4 6	7 9 11	13 15 17
·91	8128	8147	8166	8185	8204	8222	8241	8260	8279	8299	2 4 6	8 9 11	13 15 17
·92	8318	8337	8356	8375	8395	8414	8433	8453	8472	8492	2 4 6	8 10 12	14 15 17
·93	8511	8531	8551	8570	8590	8610	8630	8650	8670	8690	2 4 6	8 10 12	14 16 18
·94	8710	8730	8750	8770	8790	8810	8831	8851	8872	8892	2 4 6	8 10 12	14 16 18
·95	8913	8933	8954	8974	8995	9016	9036	9057	9078	9099	2 4 6	8 10 12	15 17 19
·96	9120	9141	9162	9183	9204	9226	9247	9268	9290	9311	2 4 6	8 11 13	15 17 19
·97	9333	9354	9376	9397	9419	9441	9462	9484	9506	9528	2 4 7	9 11 13	15 17 20
·98	9550	9572	9594	9616	9638	9661	9683	9705	9727	9750	2 4 7	9 11 13	16 18 20
·99	9772	9795	9817	9840	9863	9886	9908	9931	9954	9977	2 5 7	9 11 14	16 18 20

Answers to exercises

Exercise 1

1.   (a)   8 sets
    (b)   1 set
    (c)   10 sets
    (d)   none
    (e)   11 sets
    (f)   3 sets
    (g)   8 sets
    (h)   1 set
    (i)   2 sets
    (j)   1 set
2.   (a), (b) and (c)
3.   (a)   3 x 3 x 3
    (b)   2 x 2 x 7
    (c)   2 x 2 x 3 x 7
    (d)   2 x 2 x 2 x 2 x 3 x 3
    (e)   7 x 31
4.   (a)   4
    (b)   7
    (c)   14
5.   (a)   24
    (b)   84
    (c)   48
6.   (a)   2, 72
    (b)   3, 90
    (c)   3, 135
7.   9032
8.   25324
9.   1778
10.   1654
11.   1288034
12.   695374
13.   124
14.   215

Exercise 2

1.   (a)   216
    (b)   1280
    (c)   905
    (d)   6
    (e)   227
    (f)   432
    (g)   99
    (h)   1
    (i)   396
    (j)   2
2.   (a)   65
    (b)   166
    (c)   36
    (d)   2964

Exercise 3

2.   (a) $\dfrac{5}{2}$

    (b) $\dfrac{21}{5}$

    (c) $\dfrac{311}{16}$

    (d) $\dfrac{27}{8}$

    (e) $\dfrac{93}{4}$

3.   (a) $3\dfrac{2}{5}$

    (b) $3\dfrac{2}{7}$

(c) $2\frac{4}{21}$

(d) $4\frac{5}{11}$

(e) $4\frac{17}{20}$

4   (a) $10\frac{19}{48}$

(b) $2\frac{9}{16}$

(c) $\frac{9}{28}$

(d) $1\frac{61}{64}$

5   (a) $75\frac{5}{6}$

(b) $\frac{3}{4}$

(c) $3\frac{6}{7}$

(d) $4\frac{3}{16}$

6   (a) $12\frac{11}{24}$

(b) $2\frac{244}{245}$

(c) $5\frac{1}{10}$

(d) $\frac{91}{192}$

7   60 mm, 60 mm, 40 mm and 80 mm

8   Rise 215 mm and going 173 mm

9    $\frac{1}{90}$

Exercise 4

1   (a)   21.7
(b)   14.167
(c)   2.9
(d)   0.113
(e)   0.001
(f)   3.19
(g)   2.7
(h)   4.71

2   (a)   0.328
(b)   0.667
(c)   0.625
(d)   0.25
(e)   0.4375
(f)   14.08
(g)   21.2
(h)   3.75
(i)   101.133
(j)   0.051

3   (a)   16.986
(b)   0.013
(c)   0.032
(d)   2
(e)   60
(f)   100

4   0.25
5   2.95 m^3
6   £54
7   Cement 3733 kg, sand 10 m^3
8   611.86 kg
9   11 rolls
10   (a)   113 litres
(b)   135 litres

Exercise 5

1   (a)   25%
(b)   10%
(c)   15%
(d)   4%
(e)   75%

2   (a) $\frac{1}{3}$

(b) $\frac{3}{20}$

(c) $\frac{1}{40}$

(d) $\frac{1}{5}$

(e) $\frac{3}{5}$

3   (a)   9.375% very porous
(b)   1.05% rather porous
(c)   0.6% only fair
4   £61.92

5 2.2%
6 15.3 m², £151.47
7 Copper 45 kg, tin 5 kg
8 7.7 m
9 41½ pence, £2.31
10 £6.76½
11 5896.8 kg
12 1.246 m
13 £3150
14 £17,143

Exercise 6

1 (a) 784
   (b) 10.69
   (c) 1775
   (d) 34,230
   (e) 0.7225
   (f) 150.1
2 (a) 13.15
   (b) 3.674
   (c) 49
   (d) 0.275
   (e) 3.5
   (f) 29
3 (a) 0.04348
   (b) 0.007752
   (c) 1.235
   (d) 0.0001712
   (e) 2.7
   (f) 0.009524
4 (a) 4485
   (b) 4.044
   (c) 787

Exercise 7

1 (a) 15^6
   (b) x^3
   (c) 8^3
   (d) 16^1
2 (a) 1
   (b) 3
   (c) 0
   (d) 2
   (e) 4

3 (a) 1.3979
   (b) 1.5336
   (c) 0.4734
   (d) 0.2385
   (e) 0.4972
   (f) 3.2375
   (g) 0.7959
   (h) 1.4314
   (i) 0.9375
   (j) 0.1504
   (k) 0.4048
   (l) 3.9849
   (m) 4.4140
   (n) 4.8859
4 All have the same mantissa

Exercise 8

1 (a) 835.5
   (b) 295.2
   (c) 422.5
2 (a) 3.8
   (b) 38.09
   (c) 2.158
   (d) 188.9
3 (a) 19.14
   (b) 6.146
   (c) 4.99
4 (a) 40.64
   (b) 2070
   (c) 13.48
   (d) 6.973
5 (a) 67.88
   (b) 190.8
   (c) 7.299
   (d) 1.36
6 (a) 0.7168
   (b) 3.267
   (c) 3.303
   (d) 147

Exercise 9

1 (a) $\bar{2}.8475$
   (b) $\bar{1}.2258$
   (c) $\bar{3}.9795$

	(d)	$\overline{2}.9586$
	(e)	$\overline{5}.0792$
	(f)	$\overline{2}.6335$
2	(a)	0.7437
	(b)	0.0863
	(c)	0.00005034
	(d)	0.002205
	(e)	0.1488
	(f)	5.782
3	(a)	0.5078
	(b)	1.173
	(c)	0.7598
	(d)	27.2
4	(a)	0.02792
	(b)	0.2183
	(c)	2.414
	(d)	0.06304
	(e)	0.03659
5	(a)	0.09251
	(b)	67.82
6	(a)	0.8557
	(b)	0.1364
	(c)	0.4609
	(d)	0.7044
	(e)	0.2052
	(f)	0.9995
7	(a)	0.8013
	(b)	0.0057
	(c)	0.05059

Exercise 10

1   39.400 m
2   3.155
3   2.357 m
4   2.34 m^2
5   53.41 m^2
6   £46.41
7   2816 m^2
8   10.5 m^2 (correct to one
     decimal place)
9    $66\frac{2}{3}$ m
10   10,542 m^2
11   Rectangular area by £1 per m^2
12   1.910 m
13   18 m^2, 14.08 m

14   20 m
15   12 m
16   90 mm
17   50 m
18   3 m
19   2.7 m
20   74.25 m^2
21   53 m^2
22   24 m
23   200 mm x 200 mm

Exercise 11

1   22.98 m^2
2   11.31 m
3   25,270 mm^2
4   20,560 mm^2
5   17.28 m^2, 10.6 m
6   8.908 m^2, 9.897 m
7   629,700 mm^2 (or 0.6297 m^2),
     2.08 m
8   3.64 m^2, 5.027 m
9   1.885 m, 0.2749 m^2 (or
     274,900 mm^2)
10   27.54 m^2
11   7.083 m^2, 3.097 m^2
12   (a)   221,800 mm^2, outer
            535.9 mm, inner 92.37 mm
     (b)   1.559 m^2, 5.129 m
     (c)   1.285 m^2, 7.574 m
     (d)   1862 mm^2, 157.1 mm

Exercise 12

1   (a) $43°$
     (b) $62°$
     (c) $75°$
     (d) $66°$
     (e) $82°$
2   (a) $40°$
     (b) $84°$
     (c) $127°$
     (d) $162°$
     (e) $56°$
3   B = $60°$ and A = $80°$
4   (a) $105°$, 3.844, 5.492

(b) 90° (3:4:5 triangle)

(c) $52\frac{1}{2}$° each (isosceles triangle)

5 17.67 cm

6 2.078 m height, 1.200 m out

7 (a) AG, GF, FE, GC and CF = 2.000 m

(b) BG and DF = 1.000 m

(c) AB, BC, CD, DE = 1.732 m

(d) Angles ABG, CBG, CDF, EDF = 90°

(e) Angles BCG, FCD, DEF = 30°

(f) Angles remaining = 60°

8 Height 1.9485 m, steep rafter 2.250 m, other rafter 3.897 m

9 1.6986 m

10 1.109 m²

11 1.169 m

12 60°, 20 m (pole and ropes form 60°/30° triangles)

13 4.884 m, $47\frac{1}{2}$°

Exercise 13

1 1.134 m³

2 132.16 kg

3 342 litres

4 1080 m³

5 £36.56

6 884.5 kg

7 $17\frac{3}{4}$ days

8 38.48 m³

9 0.06038 m³, 10656 m²

10 132.86 m³

11 1109 kgf

Exercise 14

1 (a) 7273 kg/m³

(b) 6.42 kg

2 2.6 m³

3 15 m³

4 2.757 kgf

5 0.3928 m³

6 51.84 m², 33.88 m³

7 (a) 3.535 litres

(b) 1119 cm²

8 (a) 697.9 m³

(b) 95.59 m²

9 288.6 mm

10 258 mm

11 8.834 m²

12 1.696 m³

13 15.23 m²

14 223.5 mm

15 5.656 litres

Exercise 15

1 (a) $1.5\,b\sqrt{h^2 - b^2}$

(b) $R^2\,(\pi - \frac{\sqrt{3}}{2})$ or 2.276 R^2

(c) $r\,(a + b + \frac{\pi r}{2})$

2 $A = lb/9,\ 4/27$

3 $W = x\,(1.5\,n + 1\frac{1}{3})$

4 11.88 m²

5 180

6 4.877 m²

7 $t = \frac{1}{5}\sqrt{d},\ 3$ mm

8 (a) $W = \frac{4}{3}\,\pi\,r^3\,w$

(b) 3.776 kg

9 185,600

10 $\frac{\pi r^2\,hwp}{500}$, £762

11 848.8 N/mm²

12 2.574 m

Exercise 16

1 (b) 9500 mm²

(c) 9507 mm²

3 37.2°C, 230°F

4 (a) 1.93

(b)   3
(c)   24.52 m^3
5   Dry concrete
6   (a)   22.56 cm^2
(b)   7.48 cm

Exercise 17

1   435 kN/m^2, 120 mm
2   2304 kg/m^3, 672 kg/m^3,
2000 kg/m^3
3   (a)   34.25 minutes
(b)   24,000 litres
(c)   142,800 litres
4   14.06 m^2, 8.66
5   202 g, 5$\frac{1}{2}$ days (nearest $\frac{1}{2}$ day)
6   39 N, 14.2 kg
7   235 kN, 3.500 m
8   (a)   2.5 kg, 3.29 kg, 4.42 kg
(b)   7.6 lb, 2.3 lb, 10.5 lb
9   5.780 m, 8.2 kN/m^2

Exercise 18

1   195.4 kN/m^2
2   3.118 N/mm^2
3   150 mm
4   39 kJ
5   70.63%
6   5297.4 Nm, 5.297 kJ, 143.4 kJ
7   No
8   40.5 Nm, 763.5 J
9   245.3 N
10   (a)   900 N (or 0.9 kN)
downwards
(b)   Any force at such a
distance from *A* which
will have a clockwise
moment of 98 kNm

Exercise 19

1   £127.65
2   21 m^2, £31.50

3   (a)   £64.52
(b)   £12.90
4   £332.16
5   £1603.80
6   £2604.45
7   £1.85
8   £6.07
9   £35.43
10   £1104.48
11   £57.80
12   £15.13 per metre
13   £3525.08
14   £352.43
15   £159.50
16   £165.70
17   £407.73
18   £161.02

Revision exercise 1

1   18 = 2 x 3 x 3
21 = 3 x 7
24 = 2 x 2 x 2 x 3
27 = 3 x 3 x 3
2   (a)   94, 4
(b)   60, 2
(c)   180, 15
3   (a) $\frac{39}{7}$
(b) $\frac{58}{5}$
(c) $\frac{135}{8}$
(d) $\frac{76}{9}$
(e) $\frac{451}{64}$
(f) $\frac{343}{15}$
(g) $\frac{1307}{128}$
(h) $\frac{515}{144}$
4   (a) $14\frac{1}{4}$
(b) $3\frac{5}{16}$

(c) $29\frac{3}{8}$

(d) $4\frac{49}{144}$

(e) $5\frac{71}{128}$

(f) $3\frac{1}{7}$

(g) $8\frac{1}{5}$

(h) $6\frac{1}{4}$

5 (a) $14\frac{1}{12}$

(b) $40\frac{37}{48}$

(c) $16\frac{1}{6}$

(d) $24\frac{9}{14}$

6 (a) $4\frac{5}{6}$

(b) $6\frac{29}{96}$

(c) $5\frac{3}{20}$

(d) $3\frac{3}{28}$

7 (a) $6\frac{1}{6}$

(b) $7\frac{29}{42}$

(c) $6\frac{7}{32}$

(d) $7\frac{44}{45}$

8 (a) $6\frac{1}{3}$

(b) $14\frac{13}{16}$

(c) 9

9 (a) $2\frac{1}{5}$

(b) 12

(c) $28\frac{1}{2}$

10 (a) 85.242
    (b) 149.839
    (c) 64.03
    (d) 28.284
    (e) 110.552
11 (a) 19.56
    (b) 47.45
12 (a) 21.62
    (b) 243.98
    (c) 2.27
    (d) 35.3
    (e) 2.07
    (f) 25.06
13 (a) 0.813
    (b) 0.273
    (c) 0.318
    (d) 0.609
14 10 mm
15 2.700 m
16 4.510 m
17 176.5 g
18 (a) £320.75
    (b) 1.435 m
    (c) 7.5 m²
    (d) 225 kg
19 0.214 m³
20 45 kg
21 78.12 kg
22 Tin 10.5 kg, lead 24.5 kg
23 £111.88
24 £54
25 15 m³

Revision exercise 2

1 (a) 529
   (b) 364.8
   (c) 61.46
   (d) 0.0625 (or $\frac{1}{16}$)

   (e) $0.\dot{4}$ (or $\frac{4}{9}$)

   (f) 3.0625 (or $3\frac{1}{16}$)
   (g) 9.87
   (h) 9
2 (a) 11
   (b) 3

(c)   0.375 (or $\frac{3}{8}$)

(d)   5

3   (a)   24

(b)   96

(c)   14

(d)   35

(e)   39

(f)   72

(g)   55

4   (a)   9

(b)   2.25

5   (a)   9378

(b)   8772

(c)   1708

(d)   1227

(e)   4095

(f)   39.8

(g)   5.166

(h)   23.14

(i)   12.47

6   (a)   1,825,000

(b)   16.31

7   (a)   0.2094

(b)   1.759

(c)   9.097

(d)   2.344

(e)   0.5039

(f)   0.4018

8   £17.56

9   33.2 m^2

10   8.79 m^2, 42.9 m

11   10.71 m^2

12   7.8 m^2

13   150 m

14   16 m, £9.12

15   611 m^2

16   (a)   9.62 m^2

(b)   1.227 m^2

(c)   4.525 m^2

(d)   101,800 mm^2

(e)   45.83 m^2

(f)   12.93 m^2

17   25.98 m^2

18   78.55 mm

19   4.873 m^2

20   6.02 m^2

21   329,900 mm^2

22   4.91 m^2

23   15.27 m^2, 14.139 m

24   2.475 m^2, 4.005 m

25   403.2 kg

26   8500 litres per minute

27   139 kg

28   12,150 kg

29   135 kg

30   95.625 m^3

31   57.05 m^3

32   78.98 m^3

33   1.386 kgf

34   (a)   15.645 m^3

(b)   21.6 m^2

(c)   6.953 m^2

35   4687.5 m^3

36   7358 litres

37   101.8 m^2

38   5.074 kgf

39   110.4 m^3, 88.35 m^2

40   870.4 kgf

41   7.938 m^3

42   17.23 m^2

43   1,178,000 mm^3 (or 0.001178 m^3)

44   2.4 m, 4.157 m

45   2.97 m

Revision exercise 3

1   165,000 mm^2

2   1.59 m^2

3    $\frac{7}{9}$

4   2116 litres

5    $A = BD - bd$

$A = t(w + x)$

$A = t^2 + t\,(a + b)$

6   0.000039 ab

7    $W = 2.24\,xyz$

8    $W = 0.44P + 0.05N$, £81.70

9   19.48%

10   (a) $11r\,(h + r) + 2rh$ or $r\,(11h + 11r + 2h)$

(b)   851.5 cm^2

11   29.32 mm

12    $p = \dfrac{(w_2 - w_1)}{w_1} \times 100\%$, 25%

13   24 units^2

15   (a)   139.24 kN/m^2
       (b)   3.5 m

16   (a)   22.9 N/mm^2
       (b)   12 days

17   12.1 kg, 1.7 m

18   (a)   283.4 g
       (b)   12.35 oz

19   (a)   £72
       (b)   £106.20
       (c)   48 hours

20   240.3 kN/m^2

21   30 N/mm^2

22   4.8 kJ

23   73.38%

24   (a)   117.8 kJ
       (b)   36.79 kJ
       (c)   31.23 kJ

25   (a)   442
       (b)   £84.86

26   24,600, £3357.90

27   £202.78

28   £70.78

29   (a)   £276.41
       (b)   430.3 kg

30   £3093.75

31   £1140.52

32   £7482.78

33   £777.84